Quantum Theory: A Very Short Introduction

'John Polkinghorne has produced an excellent piece of work. . . . Many authors of "popular" books on modern physics have the regrettable habit of mixing science fact with science fiction. Polkinghorne never does that: he always allows the truth to stand by itself and show its own fascination. . . . I think that this is an excellent contribution to the literature on quantum theory for a general audience.'
Chris Isham, Imperial College, London

'This splendid book explains both the triumph and the mystery that is quantum theory. It is a triumph because of its towering mathematical structure, and amazing empirical accuracy. It is a mystery because of the conundrums about how to interpret it.
John Polkinghorne, himself a distinguished quantum physicist, is a sure guide to all this: he celebrates the successes of the theory, and shows unfailingly good judgment about the conundrums.'
Jeremy Butterfield, Oxford University

VERY SHORT INTRODUCTIONS are for anyone wanting a stimulating and accessible way into a new subject. They are written by experts, and have been translated into more than 45 different languages.

The series began in 1995, and now covers a wide variety of topics in every discipline. The VSI library now contains over 500 volumes—a Very Short Introduction to everything from Psychology and Philosophy of Science to American History and Relativity—and continues to grow in every subject area.

Titles in the series include the following:

John Polkinghorne

QUANTUM
THEORY

A Very Short Introduction

OXFORD
UNIVERSITY PRESS

OXFORD

UNIVERSITY PRESS

Great Clarendon Street, Oxford OX2 6DP

Oxford University Press is a department of the University of Oxford.
It furthers the University's objective of excellence in research, scholarship,
and education by publishing worldwide in

Oxford New York

Auckland Bangkok Buenos Aires Cape Town Chennai
Dar es Salaam Delhi Hong Kong Istanbul Karachi Kolkata
Kuala Lumpur Madrid Melbourne Mexico City Mumbai Nairobi
São Paulo Shanghai Taipei Tokyo Toronto

Oxford is a registered trade mark of Oxford University Press
in the UK and in certain other countries

Published in the United States
by Oxford University Press Inc., New York

© John Polkinghorne 2002

The moral rights of the author have been asserted

Database right Oxford University Press (maker)

First published as a Very Short Introduction 2002

British Library Cataloguing in Publication Data
Data available

Library of Congress Cataloging in Publication Data
Data available

ISBN 978-0-19-280252-1

22

Typeset by RefineCatch Ltd, Bungay, Suffolk
Printed in Great Britain by
Ashford Colour Press Ltd, Gosport, Hants.

To the memory of
Paul Adrien Maurice Dirac
1902–1984

I think I can safely say that no one understands quantum mechanics

Richard Feynman

Acknowledgements

I am grateful to the staff of Oxford University Press for their help in preparing the manuscript for press, and particularly to Shelley Cox for a number of helpful comments on the first draft.

Queens' College   John Polkinghorne
Cambridge

Contents

Preface

The discovery of modern quantum theory in the mid-1920s brought about the greatest revision in our thinking about the nature of the physical world since the days of Isaac Newton. What had been considered to be the arena of clear and determinate process was found to be, at its subatomic roots, cloudy and fitful in its behaviour. Compared with this revolutionary change, the great discoveries of special and general relativity seem not much more than interesting variations on classical themes. Indeed, Albert Einstein, who had been the progenitor of relativity theory, found modern quantum mechanics so little to his metaphysical taste that he remained implacably opposed to it right to the end of his life. It is no exaggeration to regard quantum theory as being one of the most outstanding intellectual achievements of the 20th century and its discovery as constituting a real revolution in our understanding of physical process.

That being so, the enjoyment of quantum ideas should not be the sole preserve of theoretical physicists. Although the full articulation of the theory requires the use of its natural language, mathematics, many of its basic concepts can be made accessible to the general reader who is prepared to take a little trouble in following through a tale of remarkable discovery. This little book is written with such a reader in mind. Its main text does not contain any mathematical equations at all. A short appendix outlines some simple mathematical insights that will give extra illumination to those able to stomach somewhat stronger meat. (Relevant sections of this appendix are cross-referenced in bold type in the main text.)

Quantum theory has proved to be fantastically fruitful during the more than 75 years of its exploitation following the originating discoveries. It is currently applied with confidence and success to the discussion of quarks and gluons (the contemporary candidates for the basic constituents of nuclear matter), despite the fact that these entities are at least 100 million times smaller than the atoms whose behaviour was the concern of the quantum pioneers. Yet a profound paradox remains. The epigraph of this book has about it some of the exuberant exaggeration of expression that characterized the discourse of that great second-generation quantum physicist, Richard Feynman, but it is certainly the case that, though we know how to do the sums, we do not *understand* the theory as fully as we should. We shall see in what follows that important interpretative issues remain unresolved. They will demand for their eventual settlement not only physical insight but also metaphysical decision.

As a young man I had the privilege of learning my quantum theory at the feet of Paul Dirac, as he gave his celebrated Cambridge lecture course. The material of Dirac's lectures corresponded closely to the treatment given in his seminal book, *The Principles of Quantum Mechanics*, one of the true classics of 20th-century scientific publishing. Not only was Dirac the greatest theoretical physicist known to me personally, his purity of spirit and modesty of demeanour (he never emphasized in the slightest degree his own immense contributions to the fundamentals of the subject) made him an inspiring figure and a kind of scientific saint. I humbly dedicate this book to his memory.

List of illustrations

Chapter 1
Classical cracks

The first flowering of modern physical science reached its culmination in 1687 with the publication of Isaac Newton's *Principia*. Thereafter mechanics was established as a mature discipline, capable of describing the motions of particles in ways that were clear and deterministic. So complete did this new science seem to be that, by the end of the 18th century, the greatest of Newton's successors, Pierre Simon Laplace, could make his celebrated assertion that a being, equipped with unlimited calculating powers and given complete knowledge of the dispositions of all particles at some instant of time, could use Newton's equations to predict the future, and to retrodict with equal certainty the past, of the whole universe. In fact, this rather chilling mechanistic claim always had a strong suspicion of hubris about it. For one thing, human beings do not experience themselves as being clockwork automata. And for another thing, imposing as Newton's achievements undoubtedly were, they did not embrace all aspects of the physical world that were then known. There remained unsettled issues that threatened belief in the total self-sufficiency of the Newtonian synthesis. For example, there was the question of the true nature and origin of the universal inverse-square law of gravity that Sir Isaac had discovered. This was a matter about which Newton himself had declined to frame a hypothesis. Then there was the unresolved question of the nature of light. Here Newton did permit himself a degree of speculative latitude. In the *Opticks* he inclined to the view that a beam of light was made up of a stream of tiny particles. This kind of corpuscular theory was consonant with Newton's tendency to view the physical world in atomistic terms.

The nature of light

It turned out that it was not until the 19th century that there was real progress in gaining understanding of the nature of light. Right at the century's beginning, in 1801, Thomas Young presented very convincing evidence for the fact that light had the character of a wave motion, a speculation that had been made more than a century earlier by Newton's Dutch contemporary Christiaan Huygens. The key observations made by Young centred on effects that we now call interference phenomena. A typical example is the existence of alternating bands of light and darkness, which, ironically enough, had been exhibited by Sir Isaac himself in a phenomenon called Newton's rings. Effects of this kind are characteristic of waves and they arise in the following way. The manner in which two trains of waves combine depends upon how their oscillations relate to each other. If they are in step (in phase, the physicists say), then crest coincides constructively with crest, giving maximum mutual reinforcement. Where this happens in the case of light, one gets bands of brightness. If, however, the two sets of waves are exactly out of step (out of phase), then crest coincides with trough in mutually destructive cancellation, and one gets a band of darkness. Thus the appearance of interference patterns of alternating light and dark is an unmistakable signature of the presence of waves. Young's observations appeared to have settled the matter. Light is wavelike.

As the 19th century proceeded, the nature of the wave motion associated with light seemed to become clear. Important discoveries by Hans Christian Oersted and by Michael Faraday showed that electricity and magnetism, phenomena that at first sight seemed

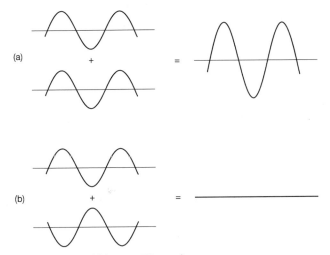

1. Adding waves: (a) in step; (b) out of step

very different in their characters, were, in fact, intimately linked
with each other. The way in which they could be combined to give a
consistent theory of electromagnetism was eventually determined
by James Clerk Maxwell – a man of such genius that he can fittingly
be spoken of in the same breath as Isaac Newton himself. Maxwell's
celebrated equations, still the fundamental basis of electromagnetic
theory, were set out in 1873 in his *Treatise on Electricity and
Magnetism*, one of the all-time classics of scientific publishing.
Maxwell realized that these equations possessed wavelike solutions
and that the velocity of these waves was determined in terms of
known physical constants. This turned out to be the known velocity
of light!

This discovery has been regarded as the greatest triumph of
19th-century physics. The fact that light was electromagnetic waves
seemed as firmly established as it could possibly be. Maxwell and
his contemporaries regarded these waves as being oscillations in an
all-pervading elastic medium, which came to be called ether. In an

encyclopedia article, he was to say that the ether was the best confirmed entity in the whole of physical theory.

We call the physics of Newton and Maxwell classical physics. By the end of the 19th century it had become an imposing theoretical edifice. It was scarcely surprising that grand old men, like Lord Kelvin, came to think that all the big ideas of physics were now known and all that remained to do was tidy up the details with increased accuracy. In the last quarter of the century, a young man in Germany contemplating an academic career was warned against going into physics. It would be better to look elsewhere, for physics was at the end of the road, with so little really worthwhile left to do. The young man's name was Max Planck, and fortunately he ignored the advice he had been given.

As a matter of fact, some cracks had already begun to show in the splendid facade of classical physics. In the 1880s, the Americans Michelson and Morley had done some clever experiments intended to demonstrate the Earth's motion through the ether. The idea was that, if light was indeed waves in this medium, then its measured speed should depend upon how the observer was moving with respect to the ether. Think about waves on the sea. Their apparent velocity as observed from a ship depends upon whether the vessel is moving with the waves or against them, appearing less in the former case than in the latter. The experiment was designed to compare the speed of light in two mutually perpendicular directions. Only if the Earth were coincidently at rest with respect to the ether at the time at which the measurements were made would the two speeds be expected to be the same, and this possibility could be excluded by repeating the experiment a few months later, when the Earth would be moving in a different direction in its orbit. In fact, Michelson and Morley could detect no difference in velocity. Resolution of this problem would require Einstein's special theory of relativity, which dispensed with an ether altogether. That great discovery is not the concern of our present story, though one should note that relativity, highly significant and

4

surprising as it was, did not abolish the qualities of clarity and determinism that classical physics possessed. That is why, in the Preface, I asserted that special relativity required much less by way of radical rethinking than quantum theory was to demand.

Spectra

The first hint of the quantum revolution, unrecognized as such at the time, actually came in 1885. It arose from the mathematical doodlings of a Swiss schoolmaster called Balmer. He was thinking about the spectrum of hydrogen, that is to say the set of separated coloured lines that are found when light from the incandescent gas is split up by being passed through a prism. The different colours correspond to different frequencies (rates of oscillation) of the light waves involved. By fiddling around with the numbers, Balmer discovered that these frequencies could be described by a rather simple mathematical formula [see Mathematical appendix, 1]. At the time, this would have seemed little more than a curiosity.

Later, people tried to understand Balmer's result in terms of their contemporary picture of the atom. In 1897, J. J. Thomson had discovered that the negative charge in an atom was carried by tiny particles, which eventually were given the name 'electrons'. It was supposed that the balancing positive charge was simply spread throughout the atom. This idea was called 'the plum pudding model', with the electrons playing the role of the plums and the positive charge that of the pudding. The spectral frequencies should then correspond to the various ways in which the electrons might oscillate within the positively-charged 'pudding'. It turned out, however, to be extremely difficult to make this idea actually work in an empirically satisfactory way. We shall see that the true explanation of Balmer's odd discovery was eventually to be found using a very different set of ideas. In the meantime, the nature of atoms probably seemed too obscure a matter for these problems to give rise to widespread anxiety.

The ultraviolet catastrophe

Much more obviously challenging and perplexing was another difficulty, first brought to light by Lord Rayleigh in 1900, which came to be called 'the ultraviolet catastrophe'. It had arisen from applying the ideas of another great discovery of the 19th century, statistical physics. Here scientists were attempting to get to grips with the behaviour of very complicated systems, which had a great many different forms that their detailed motions could take. An example of such a system would be a gas made up of very many different molecules, each with its own state of motion. Another example would be radiative energy, which might be made up of contributions distributed between many different frequencies. It would be quite impossible to keep track of all the detail of what was happening in systems of this complexity, but nevertheless some important aspects of their overall behaviour could be worked out. This was the case because the bulk behaviour results from a coarse-grained averaging over contributions from many individual component states of motion. Among these possibilities, the most probable set dominates because it turns out to be overwhelmingly most probable. On this basis of maximizing likelihood, Clerk Maxwell and Ludwig Bolzmann were able to show that one can reliably calculate certain bulk properties of the overall behaviour of a complex system, such as the pressure in a gas of given volume and temperature.

Rayleigh applied these techniques of statistical physics to the problem of how energy is distributed among the different frequencies in the case of black body radiation. A black body is one that perfectly absorbs all the radiation falling on it and then re-emits all of that radiation. The issue of the spectrum of radiation in equilibrium with a black body might seem a rather exotic kind of question to raise but, in fact, there are excellent approximations to black bodies available, so this is a matter that can be investigated experimentally as well as theoretically, for example by studying radiation in the interior of a specially prepared oven. The question was simplified by the fact that it was known that the answer should depend only on the temperature of the body and not on any further details of its structure. Rayleigh pointed out that straightforward application of the well-tried ideas of statistical physics led to a disastrous result. Not only did the calculation not agree with the measured spectrum, but it did not make any sense at all. It predicted that an infinite amount of energy would be found concentrated in the very highest frequencies, an embarrassing conclusion that came to be called 'the ultraviolet catastrophe'. The catastrophic nature of this conclusion is clear enough: 'ultraviolet' is then a way of saying 'high frequencies'. The disaster arose because classical statistical physics predicts that each degree of freedom of the system (in this case, each distinct way in which the radiation can wave) will receive the same fixed amount of energy, a quantity that depends only on the temperature. The higher the frequency, the greater the number of corresponding modes of oscillation there are, with the result that the very highest frequencies run away with everything, piling up unlimited quantities of energy. Here was a problem that amounted to rather more than an unsightly flaw on the face of the splendid facade of classical physics. It was rather the case of a gaping hole in the building.

Within a year, Max Planck, now a professor of physics in Berlin, had found a remarkable way out of the dilemma. He told his son that he believed he had made a discovery of equal significance to those of Newton. It might have seemed a grandiose claim to make, but Planck was simply speaking the sober truth.

Classical physics considered that radiation oozed continuously in and out of the black body, much as water might ooze in and out of a sponge. In the smoothly changing world of classical physics, no other supposition seemed at all plausible. Yet Planck made a contrary proposal, suggesting that radiation was emitted or absorbed from time to time in packets of energy of a definite size. He specified that the energy content of one of these *quanta* (as the packets were called) would be proportional to the frequency of the radiation. The constant of proportionality was taken to be a universal constant of nature, now known as Planck's constant. It is denoted by the symbol h. The magnitude of h is very small in terms of sizes corresponding to everyday experience. That was why this punctuated behaviour of radiation had not been noticed before; a row of small dots very close together looks like a solid line.

An immediate consequence of this daring hypothesis was that high-frequency radiation could only be emitted or absorbed in events involving a single quantum of significally high energy. This large energy tariff meant that these high-frequency events would be severely suppressed in comparison with the expectations of classical physics. Taming of the high frequencies in this way not only removed the ultraviolet catastrophe, it also yielded a formula in detailed agreement with the empirical result.

Planck was obviously on to something of great significance. But exactly what that significance was, neither he nor others were sure about at first. How seriously should one take the quanta? Were they a persistent feature of radiation or simply an aspect of the way that radiation happened to interact with a black body? After all, drips from a tap form a sequence of aqueous quanta, but they merge with the rest of the water and lose their individual identity as soon as they fall into the basin.

The photoelectric effect

The next advance was made by a young man with time on his hands as he worked as a third-class examiner in the Patent Office in Berne. His name was Albert Einstein. In 1905, an *annus mirabilis* for Einstein, he made three fundamental discoveries. One of them proved to be the next step in the unfolding story of quantum theory. Einstein thought about the puzzling properties that had come to light from investigations into the photoelectric effect [2]. This is the phenomenon in which a beam of light ejects electrons from within a metal. Metals contain electrons that are able to move around within their interior (their flow is what generates an electric current), but which do not have enough energy to escape from the metal entirely. That the photoelectric effect happened was not at all surprising. The radiation transfers energy to electrons trapped inside the metal and, if the gain is sufficient, an electron can then escape from the forces that constrain it. On a classical way of thinking, the electrons would be agitated by the 'swell' of the light waves and some could be sufficiently disturbed to shake loose from the metal. According to this picture, the degree to which this happened would be expected to depend upon the intensity of the beam, since this determined its energy content, but one would not anticipate any particular dependence on the frequency of the incident light. In actual fact, the experiments showed exactly the reverse behaviour. Below a certain critical frequency, no electrons were emitted, however intense the beam might be; above that frequency, even a weak beam could eject some electrons.

Einstein saw that this puzzling behaviour became instantly intelligible if one considered the beam of light as a stream of persisting quanta. An electron would be ejected because one of these quanta had collided with it and given up all its energy. The amount of energy in that quantum, according to Planck, was directly proportional to the frequency. If the frequency were too low, there would not be enough energy transferred in a collision to enable the electron to escape. On the other hand, if the frequency exceeded a certain critical value, there would be enough energy for the electron to be able to get away. The intensity of the beam simply determined how many quanta it contained, and so how many electrons were involved in collisions and ejected. Increasing the intensity could not alter the energy transferred in a single collision. Taking seriously the existence of quanta of light (they came to be called 'photons'), explained the mystery of the photoelectric effect. The young Einstein had made a capital discovery. In fact, eventually he was awarded his Nobel Prize for it, the Swedish Academy presumably considering his two other great discoveries of 1905 – special relativity and a convincing demonstration of the reality of molecules – as still being too speculative to be rewarded in this fashion!

The quantum analysis of the photoelectric effect was a great physics victory, but it seemed nevertheless to be a pyrrhic victory. The subject now faced a severe crisis. How could all those great 19th-century insights into the wave nature of light be reconciled with these new ideas? After all, a wave is a spread-out, flappy thing, while a quantum is particlelike, a kind of little bullet. How could both possibly be true? For a long while physicists just had to live with the uncomfortable paradox of the wave/particle nature of light. No progress would have been made by trying to deny the insights either of Young and Maxwell or of Planck and Einstein. People just had to hang on to experience by the skin of their intellectual teeth, even if they could not make sense of it. Many seem to have done so by the rather cowardly tactic of averting their gaze. Eventually, however, we shall find that the story had a happy ending.

The nuclear atom

In the meantime, attention turned from light to atoms. In Manchester in 1911, Ernest Rutherford and some younger co-workers began to study how some small, positively charged projectiles called a-particles behaved when they impinged on a thin gold film. Many a-particles passed through little affected but, to the great surprise of the investigators, some were substantially deflected. Rutherford said later that it was as astonishing as if a 15″ naval shell had recoiled on striking a sheet of tissue paper. The plum pudding model of the atom could make no sense at all of this result. The a-particles should have sailed through like a bullet through cake. Rutherford quickly saw that there was only one way out of the dilemma. The positive charge of the gold atoms, which would repel the positive a-particles, could not be spread out as in a 'pudding' but must all be concentrated at the centre of the atom. A close encounter with such concentrated charge would be able substantially to deflect an α-particle. Getting out an old mechanics textbook from his undergraduate days in New Zealand, Rutherford – who was a wonderful experimental physicist but no great shakes as a mathematician – was able to show that this idea, of a central positive charge in the atom orbited by negative electrons, perfectly fitted the observed behaviour. The plum pudding model instantly gave way to the 'solar system' model of the atom. Rutherford and his colleagues had discovered the atomic nucleus.

Here was a great success, but it seemed at first sight to be yet another pyrrhic victory. In fact, the discovery of the nucleus plunged classical physics into its deepest crisis yet. If the electrons in an atom are encircling the nucleus, they are continually changing their direction of motion. Classical electromagnetic theory then requires that in this process they should radiate away some of their energy. As a result they should steadily move in nearer to the nucleus. This is a truly disastrous conclusion, for it implies that atoms would be unstable, as their component electrons spiralled into collapse towards the centre. Moreover, in the course of this decay, a continuous pattern of radiation would be emitted that looked nothing like the sharp spectral frequencies of the Balmer formula. After 1911, the grand edifice of classical physics was not just beginning to crack. It looked as though an earthquake had struck it.

The Bohr atom

However, as with the case of Planck and the ultraviolet catastrophe, there was a theoretical physicist at hand to come to the rescue and to snatch success from the jaws of failure by proposing a daring and radical new hypothesis. This time it was a young Dane called Niels Bohr, who was working in Rutherford's Manchester. In 1913, Bohr made a revolutionary proposal [3]. Planck had replaced the classical idea of a smooth process in which energy oozed in and out of a black body by the notion of a punctuated process in which the energy is emitted or absorbed as quanta. In mathematical terms, this meant that a quantity such as the energy exchanged, which previously had been thought of as taking any possible value, was now considered to be able only to take a series of sharp values (1, 2, 3, . . . packets involved). Mathematicians would say that the continuous had been replaced by the discrete. Bohr saw that this might be a very general tendency in the new kind of physics that was slowly coming to birth. He applied to atoms similar principles to those that Planck had applied to radiation. A classical physicist would have supposed that electrons encircling a nucleus could do so in orbits whose radii could take any value. Bohr proposed the replacement of this continuous possibility by the discrete requirement that the radii could only take a series of distinct values that one could enumerate (first, second, third, . . .). He also made a definite suggestion of how these possible radii were determined, using a prescription that involved Planck's constant, h. (The proposal related to angular momentum, a measure of the electron's rotatory motion that is measured in the same physical units as h.)

Two consequences followed from these proposals. One was the highly desirable property of re-establishing the stability of atoms. Once an electron was in the state corresponding to the lowest permitted radius (which was also the state of lowest energy), it had nowhere else to go and so no more energy could be lost. The electron might have got to this lowest state by losing energy as it moved from a state of higher radius. Bohr assumed that when this happened the surplus energy would be radiated away as a single photon. Doing the sums showed that this idea led straightforwardly to the second consequence of Bohr's bold surmise, the prediction of the Balmer formula for spectral lines. After almost 30 years, this mysterious numerical prescription changed from being an inexplicable oddity into being an intelligible property of the new theory of atoms. The sharpness of spectral lines was seen as a reflection of the discreteness that was beginning to be recognized as a characteristic feature of quantum thinking. The continuously spiralling motion that would have been expected on the basis of classical physics had been replaced by a sharply discontinuous quantum jump from an orbit of one permitted radius to an orbit of a lower permitted radius.

The Bohr atom was a great triumph. But it had arisen from an act of inspired tinkering with what was still, in many respects, classical physics. Bohr's pioneering work was, in reality, a substantial repair, patched on to the shattered edifice of classical physics. Attempts to extend these concepts further soon began to run into difficulties and to encounter inconsistencies. The 'old quantum theory', as these efforts came to be called, was an uneasy and unreconciled combination of the classical ideas of Newton and Maxwell with the quantum prescriptions of Planck and Einstein. Bohr's work was a vital step in the unfolding history of quantum physics, but it could be no more than a staging post on the way to the 'new quantum theory', a fully integrated and consistent account of these strange ideas. Before that was attained, there was another important phenomenon to be discovered that further emphasized the unavoidable necessity of finding a way to cope with quantum thinking.

Compton scattering

In 1923, the American physicist Arthur Compton investigated the scattering of X-rays (high-frequency electromagnetic radiation) by matter. He found that the scattered radiation had its frequency changed. On a wave picture, this could not be understood. The latter implied that the scattering process would be due to electrons in the atoms absorbing and re-emitting energy from the incident waves, and that this would take place without a change of frequency. On a photon picture, however, the result could easily be understood. What would be involved would be a 'billiard ball' collision between an electron and a photon, in the course of which the photon would lose some of its energy to the electron. According to the Planck prescription, change of energy is the same as change of frequency. Compton was thus able to give a quantitative explanation of his observations, thereby providing the most persuasive evidence to date for the particlelike character of electromagnetic radiation.

The perplexities to which the sequence of discoveries discussed in this chapter gave rise were not to continue unaddressed for much longer. Within two years of Compton's work, theoretical progress of a substantial and lasting kind came to be made. The light of the new quantum theory began to dawn.

Chapter 2
The light dawns

The years following Max Planck's pioneering proposal were a time of confusion and darkness for the physics community. Light was waves; light was particles. Tantalizingly successful models, such as the Bohr atom, held out the promise that a new physical theory was in the offing, but the imperfect imposition of these quantum patches on the battered ruins of classical physics showed that more insight was needed before a consistent picture could emerge. When eventually the light did dawn, it did so with all the suddenness of a tropical sunrise.

In the years 1925 and 1926 modern quantum theory came into being fully fledged. These *anni mirabiles* remain an episode of great significance in the folk memory of the theoretical physics community, still recalled with awe despite the fact that living memory no longer has access to those heroic times. When there are contemporary stirrings in fundamental aspects of physical theory, people may be heard to say, 'I have the feeling that it is 1925 all over again'. There is a wistful note present in such a remark. As Wordsworth said about the French Revolution, 'Bliss it was in that dawn to be alive, but to be young was very heaven!' In fact, though many important advances have been made in the last 75 years, there has not yet been a second time when radical revision of physical principles has been necessary on the scale that attended the birth of quantum theory.

P. DEBYE   A. PICCARD   E. HENRIOT   P. EHRENFEST   Ed. HERZEN   Th. DE DONDER   E. SCHRÖDINGER   E. VERSCHAFFELT   W. PAULI   W. HEISENBERG   R.H. FOWLER   L. BRILLOUIN

M. KNUDSEN   W.L. BRAGG   H.A. KRAMERS   P.A.M. DIRAC   A.H. COMPTON   L. de BROGLIE   M. BORN   N. BOHR

I. LANGMUIR   M. PLANCK   Mme CURIE   H.A. LORENTZ   A. EINSTEIN   P. LANGEVIN   Ch.E. GUYE   C.T.R. WILSON   O.W. RICHARDSON

Absents : Sir W.H. BRAGG, H. DESLANDRES et E. VAN AUBEL

2. **The great and the good of quantum theory: Solvay Conference 1927**

Two men in particular set the quantum revolution underway, producing almost simultaneously startling new ideas.

Matrix mechanics

One of them was a young German theorist, Werner Heisenberg. He had been struggling to understand the details of atomic spectra. Spectroscopy has played a very important role in the development of modern physics. One reason has been that experimental techniques for the measurement of the frequencies of spectral lines are capable of great refinement, so that they yield very accurate results that pose very precise problems for theorists to attack. We have already seen a simple example of this in the case of the hydrogen spectrum, with Balmer's formula and Bohr's explanation of it in terms of his atomic model. Matters had become more complicated since then, and Heisenberg was concerned with a much wider and more ambitious assault on spectral properties generally. While recuperating on the North Sea island of Heligoland from a severe attack of hay fever, he made his big breakthrough. The calculations looked pretty complicated but, when the mathematical dust settled, it became apparent that what had been involved was the manipulation of mathematical entities called matrices (arrays of numbers that multiply together in a particular way). Hence Heisenberg's discovery came to be known as matrix mechanics. The underlying ideas will reappear a little later in a yet more general form. For the present, let us just note that matrices differ from simple numbers in that, in general, they do not commute. That is to say, if A and B are two matrices, the product AB and the product BA are not usually the same. The order of multiplication matters, in contrast to numbers, where 2 times 3 and 3 times 2 are both 6. It turned out that this mathematical property of matrices has an important physical significance connected with what quantities could simultaneously be measured in quantum mechanics. [See **4** for a further mathematical generalization that proved necessary for the full development of quantum theory.]

In 1925 matrices were as mathematically exotic to the average theoretical physicist as they may be today to the average non-mathematical reader of this book. Much more familiar to the physicists of the time was the mathematics associated with wave motion (involving partial differential equations). This used techniques that were standard in classical physics of the kind that Maxwell had developed. Hard on the heels of Heisenberg's discovery came a very different-looking version of quantum theory, based on the much more friendly mathematics of wave equations.

Wave mechanics

Appropriately enough, this second account of quantum theory was called wave mechanics. Although its fully developed version was discovered by the Austrian physicist Erwin Schrödinger, a move in the right direction had been made a little earlier in the work of a young French aristocrat, Prince Louis de Broglie [5]. The latter made the bold suggestion that if undulating light also showed particlelike properties, perhaps correspondingly one should expect particles such as electrons to manifest wavelike properties. De Broglie could cast this idea into a quantitative form by generalizing the Planck formula. The latter had made the particlelike property of energy proportional to the wavelike property of frequency. De Broglie suggested that another particlelike property, momentum (a significant physical quantity, well-defined and roughly corresponding to the quantity of persistent motion possessed by a particle), should analogously be related to another wavelike property, wavelength, with Planck's universal constant again the relevant constant of proportionality. These equivalences provided a kind of mini-dictionary for translating from particles to waves, and vice versa. In 1924, de Broglie laid out these ideas in his doctoral dissertation. The authorities at the University of Paris felt pretty suspicious of such heterodox notions, but fortunately they consulted Einstein on the side. He recognized the young man's genius and the degree was awarded. Within a few years, experiments by Davisson and Germer in the United States, and by

George Thomson in England, were able to demonstrate the existence of interference patterns when a beam of electrons interacted with a crystal lattice, thereby confirming that electrons did indeed manifest wavelike behaviour. Louis de Broglie was awarded the Nobel Prize for physics in 1929. (George Thomson was the son of J. J. Thomson. It has often been remarked that the father won his Nobel Prize for showing that the electron is a particle, while the son won his Nobel Prize for showing that the electron is a wave.)

The ideas that de Broglie had developed were based on discussing the properties of freely moving particles. To attain a full dynamical theory, a further generalization would be required that allowed the incorporation of interactions into its account. This is the problem that Schrödinger succeeded in solving. Early in 1926 he published the famous equation that now goes by his name [6]. He had been led to its discovery by exploiting an analogy drawn from optics.

Although physicists in the 19th century thought of light as consisting of waves, they did not always use the full-blown calculational techniques of wave motion to work out what was happening. If the wavelength of the light was small compared to the dimensions defining the problem, it was possible to employ an altogether simpler method. This was the approach of geometrical optics, which treated light as moving in straight line rays which were reflected or refracted according to simple rules. School physics calculations of elementary lens and mirror systems are performed today in just the same fashion, without the calculators having to worry at all about the complexities of a wave equation. The simplicity of ray optics applied to light is similar to the simplicity of drawing trajectories in particle mechanics. If the latter were to prove to be only an approximation to an underlying wave mechanics, Schrödinger argued that this wave mechanics might be discoverable by reversing the kind of considerations that had led from wave optics to geometrical optics. In this way he discovered the Schrödinger equation.

Schrödinger published his ideas only a few months after Heisenberg had presented his theory of matrix mechanics to the physics community. At the time, Schrödinger was 38, providing an outstanding counterexample to the assertion, sometimes made by non-scientists, that theoretical physicists do their really original work before they are 25. The Schrödinger equation is the fundamental dynamical equation of quantum theory. It is a fairly straightforward type of partial differential equation, of a kind that was familiar to physicists at that time and for which they possessed a formidable battery of mathematical solution techniques. It was much easier to use than Heisenberg's new-fangled matrix methods. At once people could set to work applying these ideas to a variety of specific physical problems. Schrödinger himself was able to derive from his equation the Balmer formula for the hydrogen spectrum. This calculation showed both how near and yet how far from the truth Bohr had been in the inspired tinkering of the old quantum theory. (Angular momentum was important, but not exactly in the way that Bohr had proposed.)

Quantum mechanics

It was clear that Heisenberg and Schrödinger had made impressive advances. Yet at first sight the way in which they had presented their new ideas appeared so different that it was not clear whether they had made the same discovery, differently expressed, or whether there were two rival proposals on the table [see the discussion of **10**]. Important clarificatory work immediately followed, to which Max Born in Göttingen and Paul Dirac in Cambridge were particularly significant contributors. It soon became established that there was a single theory that was based on common general principles, whose mathematical articulation could take a variety of equivalent forms. These general principles were eventually most transparently set out in Dirac's *Principles of Quantum Mechanics*, first published in 1930 and one of the intellectual classics of the 20th century. The preface to the first edition begins with the deceptively simple statement, 'The methods of progress in theoretical physics have undergone a vast change during the present century'. We must now consider the transformed picture of the nature of the physical world that this vast change had introduced.

I learned my quantum mechanics straight from the horse's mouth, so to speak. That is to say, I attended the famous course of lectures on quantum theory that Dirac gave in Cambridge over a period of more than 30 years. The audience included not only final year undergraduates like myself, but frequently also senior visitors who rightly thought it would be a privilege to hear again the story, however familiar it might be to them in outline, from the lips of the man who had been one of its outstanding protagonists. The lectures followed closely the pattern of Dirac's book. An impressive feature was an utter lack of emphasis on the part of the lecturer on what had been his own considerable personal contribution to these great discoveries. I have already spoken of Dirac as a kind of scientific saint, in the purity of his mind and the singleness of his purpose. The lectures enthralled one by their clarity and the majestic unfolding of their argument, as satisfying and seemingly inevitable as the development of a Bach fugue. They were wholly free from rhetorical tricks of any kind, but near the beginning Dirac did permit himself a mildly theatrical gesture.

He took a piece of chalk and broke it in two. Placing one fragment on one side of his lectern and the other on the other side, Dirac said that classically there is a state where the piece of chalk is 'here' and one where the piece of chalk is 'there', and these are the only two possibilities. Replace the chalk, however, by an electron and in the quantum world there are not only states of 'here' and 'there' but also a whole host of other states that are mixtures of these possibilities – a bit of 'here' and a bit of 'there' added together. Quantum theory permits the mixing together of states that classically would be mutually exclusive of each other. It is this counterintuitive possibility of addition that marks off the quantum world from the everyday world of classical physics [7]. In professional jargon this new possibility is called the *superposition principle*.

Double slits and superposition

The radical consequences that follow from the assumption of superposition are well illustrated by what is called the double slits experiment. Richard Feynman, the spirited Nobel Prize physicist who has caught the popular imagination by his anecdotal books, once described this phenomenon as lying at 'the heart of quantum mechanics'. He took the view that you had to swallow quantum theory whole, without worrying about the taste or whether you could digest it. This could be done by gulping down the double slits experiment, for

> In reality it contains the *only* mystery. We cannot make the mystery go away by 'explaining' how it works. We will just *tell* you how it works. In telling you how it works we will have told you about the basic peculiarities of all quantum mechanics.

After such a trailer, the reader will surely want to get to grips with this intriguing phenomenon. The experiment involves a source of quantum entities, let us say an electron gun that fires a steady stream of particles. These particles impinge on a screen in which there are two slits, A and B. Beyond the slitted screen there is a detector screen that can register the arrival of the electrons. It could be a large photographic plate on which each incident electron will make a mark. The rate of delivery from the electron gun is adjusted so that there is only a single electron traversing the apparatus at any one time. We then observe what happens.

The electrons arrive at the detector screen one by one, and for each of them one sees a corresponding mark appearing that records its point of impact. This manifests individual electron behaviour in a particlelike mode. However, when a large number of marks have

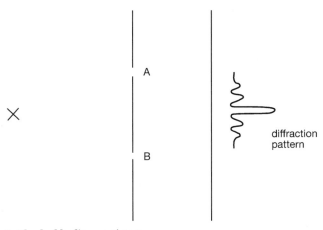

3. The double slits experiment

accumulated on the detector screen, we find that the collective pattern they have created shows the familiar form of an interference effect. There is an intense dark spot on the screen opposite the point midway between the two slits, corresponding to the location where the largest number of electron marks have been deposited. On either side of this central band there are alternating light and diminishingly dark bands, corresponding to the non-arrival and arrival of electrons at these positions respectively. Such a diffraction pattern (as the physicists call these interference effects) is the unmistakable signature of electrons behaving in a wavelike mode.

The phenomenon is a neat example of electron wave/particle duality. Electrons arriving one by one is particlelike behaviour; the resulting collective interference pattern is wavelike behaviour. But there is something much more interesting than that to be said. We can probe a little deeper into what is going on by asking the question, When an indivisible single electron is traversing the apparatus, through which slit does it pass in order to get to the detector screen? Let us suppose that it went through the top slit,

A. If that were the case, the lower slit B was really irrelevant and it might just as well have been temporarily closed up. But, with only A open, the electron would not be most likely to arrive at the midpoint of the far screen, but instead it would be most likely to end up at the point opposite A. Since this is not the case, we conclude that the electron could not have gone through A. Standing the argument on its head, we conclude that the electron could not have gone through B either. What then was happening? That great and good man, Sherlock Holmes, was fond of saying that when you have eliminated the impossible, whatever remains must have been the case, however improbable it may seem to be. Applying this Holmesian principle leads us to the conclusion that the indivisible electron *went through both slits*. In terms of classical intuition this is a nonsense conclusion. In terms of quantum theory's superposition principle, however, it makes perfect sense. The state of motion of the electron was the addition of the states (going through A) and (going through B).

The superposition principle implies two very general features of quantum theory. One is that it is no longer possible to form a clear picture of what is happening in the course of physical process. Living as we do in the (classical) everyday world, it is impossible for us to visualize an indivisible particle going through both slits. The other consequence is that it is no longer possible to predict exactly what will happen when we make an observation. Suppose we were to modify the double slits experiment by putting a detector near each of the two slits, so that it could be determined which slit an electron had passed through. It turns out that this modification of the experiment would bring about two consequences. One is that sometimes the electron would be detected near slit A and sometimes it would be detected near slit B. It would be impossible to predict where it would be found on any particular occasion but, over a long series of trials, the relative probabilities associated with the two slits would be 50–50. This illustrates the general feature that in quantum theory predictions of the results of measurement are statistical in character and not deterministic. Quantum theory deals in probabilities rather than certainties. The other consequence of this modification of the experiment would be the destruction of the interference pattern on the final screen. No longer would electrons tend to the middle point of the detector screen but they would split evenly between those arriving opposite A and those arriving opposite B. In other words, the behaviour one finds depends upon what one chooses to look for. Asking a particlelike question (which slit?) gives a particlelike answer; asking a wavelike question (only about the final accumulated pattern on the detector screen) gives a wavelike answer.

Probabilities

It was Max Born at Göttingen who first clearly emphasized the probabilistic character of quantum theory, an achievement for which he would only receive his well-deserved Nobel Prize as late as 1954. The advent of wave mechanics had raised the familiar question, Waves of what? Initially there was some disposition to suppose that it might be a question of waves of matter, so that it was the electron itself that was spread out in this wavelike way. Born soon realized that this idea did not work. It could not accommodate particlelike properties. Instead it was waves of probability that the Schrödinger equation described. This development did not please all the pioneers, for many retained strongly the deterministic instincts of classical physics. Both de Broglie and Schrödinger became disillusioned with quantum physics when presented with its probabilistic character.

The probability interpretation implied that measurements must be occasions of instantaneous and discontinuous change. If an electron was in a state with probability spread out 'here', 'there', and, perhaps, 'everywhere', when its position was measured and found to be, on this occasion, 'here', then the probability distribution had suddenly to change, becoming concentrated solely on the actually measured position, 'here'. Since the probability distribution is to be calculated from the wavefunction, this too must change discontinuously, a behaviour that the Schrödinger equation itself did not imply. This phenomenon of sudden change, called the collapse of the wavepacket, was an extra condition that had to be imposed upon the theory from without. We shall see in the next chapter that the process of measurement continues to give rise to perplexities about how to understand and interpret quantum theory. In someone like Schrödinger, the issue evoked more than perplexity. It filled him with distaste and he said that if he had known that his ideas would have led to this 'damn quantum jumping' he would not have wished to discover his equation!

Observables

(Warning to the reader: This section includes some simple mathematical ideas that are well worth the effort to acquire, but whose digestion will require some concentration. This is the only section in the main text to risk a glancing encounter with mathematics. I regret that it cannot help being somewhat hard-going for the non-mathematician.)

Classical physics describes a world that is clear and determinate. Quantum physics describes a world that is cloudy and fitful. In terms of the formalism (the mathematical expression of the theory), we have seen that these properties arise from the fact that the quantum superposition principle permits the mixing together of states that classically would be strictly immiscible. This simple principle of counterintuitive additivity finds a natural form of mathematical expression in terms of what are called vector spaces [7].

A vector in ordinary space can be thought of as an arrow, something of given length pointing in a given direction. Arrows can be added together simply by following one after the other. For example, four miles in a northerly direction followed by three miles in an easterly direction adds up to five miles in a direction 37° east of north (see

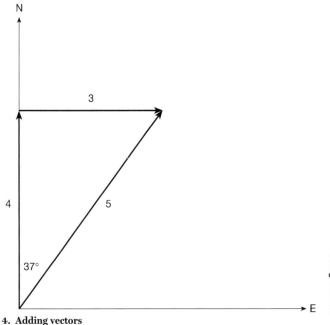

4. Adding vectors

figure 4). Mathematicians can generalize these ideas to spaces of any number of dimensions. The essential property that all vectors have is that they can be added together. Thus they provide a natural mathematical counterpart to the quantum superposition principle. The details need not concern us here but, since it is always good to feel at home with terminology, it is worth remarking that a particularly sophisticated form of vector space, called a Hilbert space, provides the mathematical vehicle of choice for quantum theory.

So far the discussion has concentrated on states of motion. One may think of them as arising from specific ways of preparing the initial material for an experiment: firing electrons from an electron gun; passing light through a particular optical system; deflecting particles by a particular set of electric and magnetic fields; and so on. One can think of the state as being 'what is the case' for the system that has been prepared, though the unpicturability of quantum theory means that this will not be as clear and straightforward a matter as it would be in classical physics. If the physicist wants to know something more precisely (where actually is the electron?), it will be necessary to make an observation, involving an experimental intervention on the system. For example, the experimenter may wish to measure some particular dynamical quantity, such as the position or the momentum of an electron. The formal question then arises: If the state is represented by a vector, how are the observables that can be measured to be represented? The answer lies in terms of operators acting on the Hilbert space. Thus the scheme linking mathematical formalism to physics includes not only the specification that vectors correspond to states, but also that operators correspond to observables [8].

The general idea of an operator is that it is something that transforms one state into another. A simple example is provided by rotation operators. In ordinary three-dimensional space, a rotation through 90° about the vertical (in the sense of a right-handed screw) turns a vector (think of it as an arrow) pointing east into a vector (arrow) pointing north. An important property of operators is that usually they do not commute with each other; that is to say the order in which they act is significant. Consider two operators: R_1, a rotation through 90° about the vertical; R_2, a rotation through 90° (again right-handed) about a horizontal axis pointing north. Apply them in the order R_1 followed by R_2 to an arrow pointing east. R_1 turns this into an arrow pointing north, which is then unchanged by R_2. We represent the two operations performed in this order as the product $R_2.R_1$, since operators, like Hebrew and Arabic, are always read from right to left. Applying the operators in the reverse order first changes the eastward arrow into an arrow pointing downwards (effect of R_2), which is then left unchanged (effect of R_1). Since $R_2.R_1$ ends up with an arrow pointing north and $R_1.R_2$

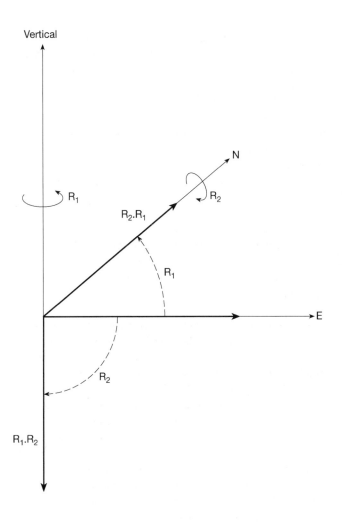

5. Non-commuting rotations

ends up with an arrow pointing downwards, these two products are quite distinct from each other. The order matters – rotations do not commute.

Mathematicians will recognize that matrices can also be considered as operators, and so the non-commutativity of the matrices that Heisenberg used is another specific example of this general operator property.

All this may seem pretty abstract, but non-commutativity proves to be the mathematical counterpart of an important physical property. To see how this comes about, one must first establish how the operator formalism for observables is related to the actual results of experiments. Operators are fairly sophisticated mathematical entities, but measurements are always expressed as unsophisticated numbers, such as 2.7 units of whatever it might be. If abstract theory is to make sense of physical observations, there must be a way of associating numbers (the results of observations) with operators (the mathematical formalism). Fortunately mathematics proves equal to this challenge. The key ideas are *eigenvectors* and *eigenvalues* [8].

Sometimes an operator acting on a vector does not change that vector's direction. An example would be a rotation about the vertical axis, which leaves a vertical vector completely unchanged. Another example would be the operation of stretching in the vertical direction. This would not change a vertical vector's direction, but it would change its length. If the stretch has a doubling effect, the length of the vertical vector gets multiplied by 2. In more general terms, we say that if an operator O turns a particular vector v into a multiple λ of itself, then v is an eigenvector of O with eigenvalue λ. The essential idea is that eigenvalues (λ) give a mathematical way of associating numbers with a particular operator (O) and a particular state (v). The general principles of quantum theory include the bold requirement that an eigenvector (also called an eigenstate) will correspond physically to a state in which measuring the observable quantity O will *with certainty* give the result λ.

A number of significant consequences flow from this rule. One is the converse, that, because there are many vectors that are not eigenvectors, there will be many states in which measuring O will not give any particular result with certainty. (Mathematica aside: it is fairly easy to see that superposing two eigenstates of O corresponding to different eigenvalues will give a state that cannot be a simple eigenstate of O.) Measuring O in states of this latter kind must, therefore, give a variety of different answers on different occasions of measurement. (The familiar probabilistic character of quantum theory is again being manifested.) Whatever result is actually obtained, the consequent state must then correspond to it; that is to say, the vector must change instantaneously to become the appropriate eigenvector of O. This is the sophisticated version of the collapse of the wavepacket.

Another important consequence relates to what measurements can be mutually compatible, that is to say, made at the same time. Suppose it is possible to measure both O_1 and O_2 simultaneously, with results λ_1 and λ_2, respectively. Doing so in one order multiplies the state vector by λ_1 and then by λ_2, while reversing the order of the observations simply reverses the order in which the λs multiplies the state vector. Since the λs are just ordinary numbers, this order does not matter. This implies that $O_2.O_1$ and $O_1.O_2$ acting on the state vector have identical effects, so that the operator order does not matter. In other words, simultaneous measurements can only be mutually compatible for observables corresponding to operators that commute with each other. Putting it the other way round, observables that do not commute will not be simultaneously measurable.

Here we see the familar cloudiness of quantum theory being manifested again. In classical physics the experimenter can measure whatever is desired whenever it is desired to do so. The physical world is laid out before the potentially all-seeing eye of the scientist. In the quantum world, by contrast, the physicist's vision is partially veiled. Our access to knowledge of quantum entities is epistemologically more limited than classical physics had supposed.

Our mathematical flirtation with vector spaces is at an end. Any reader who is dazed should simply hold on to the fact that in quantum theory only observables whose operators commute with each other can be measured simultaneously.

The uncertainty principle

What all this means was considerably clarified by Heisenberg in 1927 when he formulated his celebrated uncertainty principle. He realized that the theory should specify what it permitted to be known by way of measurement. Heisenberg's concern was not with mathematical arguments of the kind that we have just been considering, but with idealized 'thought experiments' that sought to explore the physical content of quantum mechanics. One of these thought experiments involved considering the so-called γ-ray microscope.

The idea is to find out in principle how accurately one might be able to measure the position and momentum of an electron. According to the rules of quantum mechanics, the corresponding operators do not commute. Therefore, if the theory really works, it should not be possible to know the values of both position and momentum with arbitrary accuracy. Heisenberg wanted to understand in physical terms why this was so. Let's start by trying to measure the electron's position. In principle, one way to do this would be to shine light on the electron and then look through a microscope to see where it is. (Remember these are *thought* experiments.) Now, optical instruments have a limited resolving power, which places restrictions on how accurately objects can be located. One cannot do better than the wavelength of the light being used. Of course one way to increase accuracy would be to use shorter wavelengths –

which is where the γ-rays come in, since they are very high-frequency (short wavelength) radiation. However, this ruse exacts a cost, resulting from the particlelike character of the radiation. For the electron to be seen at all, it must deflect at least one photon into the microscope. Planck's formula implies that the higher the frequency, the more energy that photon will be carrying. As a result, decreasing the wavelength subjects the electron to more and more by way of an uncontrollable disturbance of its motion through its collision with the photon. The implication is that one increasingly loses knowledge of what the electron's momentum will be after the position measurement. There is an inescapable trade-off between the increasing accuracy of position measurement and the decreasing accuracy of knowledge of momentum. This fact is the basis of the uncertainty principle: it is not possible simultaneously to have perfect knowledge of both position and momentum [**9**]. In more picturesque language, one can know where an electron is, but not know what it is doing; or one can know what it is doing, but not know where it is. In the quantum world, what the classical physicist would regard as half-knowledge is the best that we can manage.

This demi-knowledge is a quantum characteristic. Observables come in pairs that epistemologically exclude each other. An everyday example of this behaviour can be given in musical terms. It is not possible both to assign a precise instant to when a note was sounded and to know precisely what its pitch was. This is because determining the pitch of a note requires analysing the frequency of the sound and this requires listening to a note for a period lasting several oscillations before an accurate estimate can be made. It is the wave nature of sound that imposes this restriction, and if the measurement questions of quantum theory are discussed from the point of view of wave mechanics, exactly similar considerations lead back to the uncertainty principle.

There is an interesting human story behind Heisenberg's discovery. At the time he was working at the Institute in Copenhagen, whose head was Niels Bohr. Bohr loved interminable discussions and the young Heisenberg was one of his favourite conversation partners. In fact, after a while, Bohr's endless ruminations drove his younger colleague almost to distraction. Heisenberg was glad to seize the opportunity afforded by Bohr's absence on a skiing holiday to get on with his own work by completing his paper on the uncertainty principle. He then rushed it off for publication before the grand old man got back. When Bohr returned, however, he detected an error that Heisenberg had made. Fortunately the error was correctable and doing so did not affect the final result. This minor blunder involved a mistake about the resolving power of optical instruments. It so happened that Heisenberg had had trouble with this subject before. He did his doctoral work in Munich under the direction of Arnold Sommerfeld, one of the leading protagonists of the old quantum theory. Brilliant as a theorist, Heisenberg had not bothered much with the experimental work that was also supposed to be a part of his studies. Sommerfeld's experimental colleague, Wilhelm Wien, had noted this. He resented the young man's cavalier attitude and decided to put him through it at the oral examination. He stumped Heisenberg precisely with a demand to derive the resolving power of optical instruments! After the exam, Wien asserted that this lapse meant that Heisenberg should fail. Sommerfeld, of course (and rightly), argued for a pass at the highest level. In the end, there had to be a compromise and the future Nobel Prize winner was awarded his PhD, but at the lowest possible level.

Probability amplitudes

The way in which probabilities are calculated in quantum theory is in terms of what are called probability amplitudes. A full discussion would be inappropriately mathematically demanding, but there are two aspects of what is involved of which the reader should be aware. One is that these amplitudes are complex numbers, that is to say, they involve not only ordinary numbers but also i, the 'imaginary' square root of -1. In fact, complex numbers are endemic in the formalism of quantum theory. This is because they afford a very convenient way of representing an aspect of waves that was referred to in Chapter 1, in the course of discussing interference phenomena. We saw that the phase of waves relates to whether two sets of waves are in step or out of step with each other (or any possibility intermediate between these two). Mathematically, complex numbers provide a natural and convenient way of expressing these 'phase relations'. The theory has to be careful, however, to ensure that the results of observations (eigenvalues) are free from any contamination by terms involving i. This is achieved by requiring that the operators corresponding to observables satisfy a certain condition that the mathematicians call being 'hermitean' [8].

The second aspect of probability amplitudes that we need at least to be told about is that, as part of the mathematical apparatus of the theory that we have been discussing, their calculation is found to involve a combination of state vectors and observable operators. Since it is these 'matrix elements' (as such combinations are called) that carry the most direct physical significance, and because it turns out that they are formed from what one might call state-observable 'sandwiches', the time-dependence of the physics can be attributed either to a time-dependence present in the state vectors or to a time-dependence present in the observables. This observation turns out to provide the clue to how, despite their apparent differences, the theories of Heisenberg and Schrödinger do actually correspond to exactly the same physics [10]. Their seeming dissimilarity arises from Heisenberg's attributing all the time-dependence to the operators and Schrödinger's attributing it wholly to the state vectors.

The probabilities themselves, which to make sense must be positive numbers, are calculated from the amplitudes by a kind of squaring (called 'the square of the modulus') that always yields a positive number from the complex amplitude. There is also a scaling condition (called 'normalization') that ensures that when all the probabilities are added together they total up to 1 (certainly something must happen!).

Complementarity

All the while these wonderful discoveries were coming to light, Copenhagen had been the centre where assessments were made and verdicts delivered on what was happening. By this time, Niels Bohr was no longer himself making detailed contributions to technical advances. Yet he remained deeply interested in interpretative issues and he was the person to whose integrity and discernment the Young Turks, who were actually writing the pioneering papers, submitted their discoveries. Copenhagen was the court of the philosopher-king, to whom the intellectual offerings of the new breed of quantum mechanics were brought for evaluation and recognition.

In addition to this role as father-figure, Bohr did offer an insightful gloss on the new quantum theory. This took the form of his notion of complementarity. Quantum theory offered a number of alternative modes of thought. There were the alternative representations of process that could be based on measuring either all positions or all momenta; the duality between thinking of entities in terms of waves or in terms of particles. Bohr emphasized that both members of these pairs of alternatives were to be taken equally seriously, and could be so treated without contradiction because each complemented rather than conflicted with the other. This was because they corresponded to different, and mutually incompatible, experimental arrangements that could not both be employed at the same time. Either you set up a wave experiment (double slits), in which case a wavelike question was being asked that would receive a wavelike answer (an interference pattern); or you set up a particle experiment (detecting which slit the electron went through) in which case the particlelike question received a particlelike answer (two areas of impact opposite the two slits).

Complementarity was obviously a helpful idea, though it by no means resolved all interpretative problems, as the next chapter will show. As Bohr grew older he became increasingly concerned with philosophical issues. He was undoubtedly a very great physicist, but it seems to me that he was distinctly less gifted at this later avocation. His thoughts were extensive and cloudy, and many books have subsequently been written attempting to analyse them, with conclusions that have assigned to Bohr a variety of mutually incompatible philosophical positions. Perhaps he would not have been surprised at this, for he liked to say that there was a complementarity between being able to say something clearly and its being something deep and worth saying. Certainly, the relevance of complementarity to quantum theory (where the issue arises from experience and we possess an overall theoretical framework that renders it intelligible) provides no licence for the easy export of the notion to other disciplines, as if it could be invoked to 'justify' any paradoxical pairing that took one's fancy. Bohr may be thought to have got perilously close to this when he suggested that complementarity could shed light on the age-old question of determinism and free will in relation to human nature. We shall postpone further philosophical reflection until the final chapter.

Quantum logic

One might well expect quantum theory to modify in striking ways our conceptions of such physical terms as position and momentum. It is altogether more surprising that it has also affected how we think about those little logical words 'and' and 'or'.

Classical logic, as conceived of by Aristotle and the man on the Clapham omnibus, is based on the distributive law of logic. If I tell you that Bill has red hair and he is either at home or at the pub, you will expect either to find a red-haired Bill at home or a red-haired Bill at the pub. It seems a pretty harmless conclusion to draw, and formally it depends upon the Aristotelian law of the excluded middle: there is no middle term between 'at home' and 'not at home'. In the 1930s, people began to realize that matters were different in the quantum world. An electron can not only be 'here' and 'not here', but also in any number of other states that are superpositions of 'here' and 'not here'. That constitutes a middle term undreamed of by Aristotle. The consequence is that there is a special form of logic, called quantum logic, whose details were worked out by Garret Birkhoff and John von Neumann. It is sometimes called three-valued logic, because in addition to 'true' and 'false' it countenances the probabilistic answer 'maybe', an idea that philosophers have toyed with independently.

Chapter 3
Darkening perplexities

At the time at which modern quantum theory was discovered, the physical problems that held centre stage were concerned with the behaviour of atoms and of radiation. This period of initial discovery was followed in the late 1920s and early 1930s by a sustained and feverish period of exploitation, as the new ideas were applied to a wide variety of other physical phenomena. For example, we shall see a little later that quantum theory gave significant new understanding of how electrons behave inside crystalline solids. I once heard Paul Dirac speak of this period of rapid development by saying that it was a time 'when second-rate men did first-rate work'. In almost anyone else's mouth those words would have been a put-down remark of a not very agreeable kind. Not so with Dirac. All his life he had a simple and matter-of-fact way of talking, in which he said what he meant with unadorned directness. His words were simply intended to convey something of the richness of understanding that flowed from those initial fundamental insights.

This successful application of quantum ideas has continued unabated. We now use the theory equally effectively to discuss the behaviour of quarks and gluons, an impressive achievement when we recall that these constituents of nuclear matter are at least 100 million times smaller than the atoms that concerned the pioneers in the 1920s. Physicists know how to do the sums and they find that the answers continue to come out right with astonishing accuracy.

For instance, quantum electrodynamics (the theory of the interaction of electrons with photons) yields results that agree with experiment to an accuracy corresponding to an error of less than the width of a human hair in relation to the distance between Los Angeles and New York!

Considered in these terms, the quantum story is a tremendous tale of success, perhaps the greatest success story in the history of physical science. Yet a profound paradox remains. Despite the physicists' ability to do the calculations, they still do not understand the theory. Serious interpretative problems remain unresolved, and these are the subject of continuing dispute. These contentious issues concern two perplexities in particular: the significance of the probabilistic character of the theory, and the nature of the measurement process.

Probabilities

Probabilities also arise in classical physics, where their origin lies in ignorance of some of the detail of what is going on. The paradigm example is the tossing of a coin. No one doubts that Newtonian mechanics determines how it should land after being spun – there is no question of a direct intervention by Fortuna, the goddess of chance – but the motion is too sensitive to the precise and minute detail of the way the coin was tossed (details of which we are unaware) for us to be able to predict exactly what the outcome will be. We do know, however, that if the coin is fair, the odds are even, 1/2 for heads and 1/2 for tails. Similarly, for a true die the probability of any particular number ending face upwards is 1/6. If one asks for the probability of throwing either a 1 or a 2, one simply adds the separate probabilities together to give 1/3. This addition rule holds because the processes of throwing that lead to 1 or to 2 are distinct and independent of each other. Since they have no influence upon each other, one just adds the resulting odds together. It all seems pretty straightforward. Yet in the quantum world things are different.

Consider first what would be the classical equivalent of the quantum experiment with electrons and the double slits. An everyday analogue would be throwing tennis balls at a fence with two holes in it. There will be a certain probability for a ball to go through one hole and a certain probability for it to go through the other. If we are concerned with the chance that the ball lands on the other side of the fence, since it has to go through one hole or the other, we just add these two probabilities together (just as we did for the two faces of the die). In the quantum case, things are different because of the superposition principle permitting the electron to go through both slits. What classically were mutually distinct possibilities are entangled with each other quantum mechanically.

As a result, the laws for combining probabilities are different in quantum theory. If one has to sum over a number of unobserved intermediate possibilities, it is the *probability amplitudes* that have to be added together, and not the probabilities themselves. In the double slits experiment we must add the amplitude for (going through A) to the amplitude for (going through B). Recall that probabilities are calculated from amplitudes by a kind of squaring process. The effect of adding-before-squaring is to produce what a mathematician would call 'cross terms'. One can taste the flavour of this idea by considering the simple arithmetical equation

$$(2 + 3)^2 = 2^2 + 3^2 + 12$$

That 'extra' 12 is the cross term.

Perhaps that seems a little mysterious. The essential notion is as follows: In the everyday world, to get the probability of a final result you simply add together the probabilities of independent intermediate possibilities. In the quantum world, the combination of intermediate possibilities that are not directly observed takes place in a more subtle and sophisticated way. That is why the quantum calculation involves cross terms. Since the probability amplitudes are in fact complex numbers, these cross terms include phase effects, so that there can be either constructive or destructive interference, as happens in the double slits experiment.

Putting the matter in a nutshell, classical probabilities correspond to ignorance and they combine by simple addition. Quantum probabilities combine in an apparently more elusive and unpicturable way. There then arises the question: Would it, nevertheless, be possible to understand quantum probabilities as also having their origin in the physicist's ignorance of all the detail of what is going on, so that the underlying basic probabilities, corresponding to inaccessible but completely detailed knowledge of what was the case, would still add up classically?

Behind the query lies a wistful hankering on the part of some to restore determinism to physics, even if it were to prove a veiled kind of determinism. Consider, for example, the decay of a radioactive nucleus (one that is unstable and liable to break up). All that quantum theory can predict is the probability for this decay to occur. For instance, it can say that a particular nucleus has a probability of 1/2 of decaying in the next hour, but it cannot predict whether that specific nucleus will actually decay during that hour. Yet perhaps that nucleus has a little internal clock that specifies precisely when it will decay, but which we cannot read. If that were the case, and if other nuclei of the same kind had their own internal clocks whose different settings would cause them to decay at different times, then what we assigned as probabilities would arise simply from ignorance, our inability to gain access to the settings of those hidden internal clocks. Although the decays would seem random to us, they would, in fact, be completely determined by these unknowable details. In ultimate reality, quantum probability would then be no different from classical probability. Theories of this kind are called *hidden variable* interpretations of quantum mechanics. Are they in fact a possibility?

The celebrated mathematician John von Neumann believed that he had shown that the unusual properties of quantum probabilities implied that they could never be interpreted as the consequence of ignorance of hidden variables. In fact, there was an error in his argument that took years to detect. We shall see later that a deterministic interpretation of quantum theory is possible in which probabilities arise from ignorance of details. However, we shall also see that the theory that succeeds in this way has other properties that have made it seem unattractive to the majority of physicists.

Decoherence

One aspect of the problems we are considering in this chapter can be phrased in terms of asking how it can be that the quantum constituents of the physical world, such as quarks and gluons and electrons, whose behaviour is cloudy and fitful, can give rise to the macroscopic world of everyday experience, which seems so clear and reliable. An important step towards gaining some understanding of this transition has been made through a development that has taken place in the last 25 years. Physicists have come to realize that in many cases it is important to take into account, more seriously than they had done previously, the environment within which quantum processes are actually taking place.

Conventional thinking had regarded that environment as being empty, except for the quantum entities whose interactions were the subject of explicit consideration. In actual fact, this idealization does not always work, and where it does not work important consequences can flow from this fact. What had been neglected was the almost ubiquitous presence of radiation. Experiments take place in an enveloping sea of photons, some coming from the Sun and some coming from the universal cosmic background radiation that is a lingering echo from the time when the universe was about half a million years old and had just become cool enough for matter and radiation to decouple from their previous universal intermingling.

It turns out that the consequence of this virtually omnipresent background radiation is to affect the phases of the relevant probability amplitudes. Taking into account this so-called 'phase randomization' can, in certain cases, have the effect of almost entirely washing out the cross terms in quantum probability calculations. (Crudely speaking, it averages about as many pluses as minuses, giving a result near zero.) All this can occur with quite astonishing rapidity. The phenomenon is called 'decoherence'.

Decoherence has been hailed by some as providing the clue by which to understand how microscopic quantum phenomena and macroscopic classical phenomena are related to each other. Unfortunately this is only a half-truth. It can serve to make some quantum probabilities look more like classical probabilities, but it does not make them the same. There still remains the central perplexity of what is called 'the measurement problem'.

The measurement problem

In classical physics, measurement is unproblematic. It is simply the observation of what is the case. Beforehand we may be able to do no more than assign a probability of 1/2 that the coin will land heads, but if that is what we see that is simply because it is what has actually happened.

Measurement in conventional quantum theory is different because the superposition principle holds together alternative, and eventually mutually exclusive, possibilities right until the last moment, when suddenly one of them alone surfaces as the realized actuality on this occasion. We have seen that one way of thinking about this can be expressed in terms of the collapse of the wavepacket. The electron's probability was spread out over 'here', 'there', and 'everywhere', but when the physicist addresses to it the experimental question 'Where are you?' and on this particular occasion the answer 'here' turns up, then all the probability collapses onto this single actuality. The big question that has remained unanswered in our discussion so far is: How does this come about?

Measurements are a chain of correlated consequences by which a state of affairs in the microscopic quantum world produces a corresponding signal observable in the everyday world of laboratory measuring apparatus. We can clarify the point by considering a somewhat idealized, but not misleading, experiment that measures an electron's *spin*. The property of spin corresponds to electrons behaving as if they were tiny magnets. Because of an unpicturable quantum effect that the reader will simply have to be asked to take on trust, the electron's magnet can only point in two opposing directions, which we may conventionally call 'up' and 'down'.

The experiment is conducted with an initially unpolarized beam of electrons, that is to say, electrons in a state that is an even superposition of 'up' and 'down'. These electrons are made to pass through an inhomogeneous magnetic field. Because of the magnetic

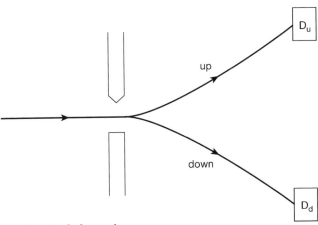

6. **A Stern-Gerlach experiment**

effect of their spin, they will be deflected either up or down according to the spin direction. They will then pass through one or other of two appropriately placed detectors, D_u or D_d (Geiger counters, perhaps), and then the experimenter will hear one or other of these detectors click, registering the passage of an electron in either the up or the down direction. This procedure is called a Stern-Gerlach experiment, after the two German physicists who first conducted an investigation of this kind. (In fact it was done with an atomic beam, but it was electrons in the atoms that controlled what was happening.) How should we analyse what is going on?

If the spin is up, the electron is deflected upwards, then it passes through D_u, the D_u clicks, and the experimenter hears D_u click. If the spin is down, the electron is deflected downwards, then it passes through D_d, the D_d clicks, and the experimenter hears this happen. One sees what is going on in this analysis. It presents us with a chain of correlated consequences: if . . . then . . . then . . . . But on an actual occasion of measurement, only one of these chains occurs. What makes this particular happening take place on this particular occasion? What settles that this time the answer shall be 'up' and not 'down'?

Decoherence does not answer this question for us. What it does do is to tighten the links in the separate chains, making them more classical-like, but it does not explain why a particular chain is the realized possibility on a particular occasion. The essence of the measurement problem is the search to understand the origin of this specificity. We shall survey the variety of responses that have been suggested, but we shall see that none of them is wholly satisfactory or free from perplexity. The proposals can be classified under a number of headings.

(1) Irrelevance

Some interpreters attempt to finesse the problem, claiming that it is irrelevant. One argument in favour of this stance is the positivist assertion that science is simply about correlating phenomena and that it should not aspire to understanding them. If we know how to do the quantum sums, and if the answers correlate highly satisfactorily with empirical experience, as they do, then that is all that we should wish for. It is simply inappropriately intellectually greedy to ask for more. A more refined form of positivism is represented by what is called 'the consistent histories' approach, which lays down prescriptions for obtaining sequences of quantum predictions that are readily interpretable as results coming from the use of classical measuring apparatus.

A different kind of argument, which also falls under the rubric of irrelevance, is the claim that quantum physics should not seek to speak about individual events at all, but its proper concern is with 'ensembles', that is to say the statistical properties of collections of events. If that were the case, a purely probabilistic account is all that one would be entitled to expect.

A third kind of argument in this general category asserts that the wavefunction is not about states of physical systems at all, but about states of the human knowledge of such systems. If one is simply thinking epistemologically, then 'collapse' is an unproblematic phenomenon: before I was ignorant; now I know. It seems very odd, however, that the representation of what is claimed to be all in the mind should actually satisfy a physical-looking equation like the Schrödinger equation.

All these arguments have a common feature. They take a minimalist view of the task of physics. In particular, they suppose that it is not concerned with gaining understanding of the detailed character of particular physical processes. This may be a view congenial to those of a certain kind of philosophical disposition, but it is abhorrent to the mind of the scientist, whose ambition is to gain the maximum attainable degree of understanding of what is happening in the physical world. To settle for less would be treason of the clerks.

(2) Large systems

The founding figures of quantum mechanics were, of course, aware of the problems that measurement posed for the theory. Niels Bohr, in particular, became very concerned with the issue. The answer that he propounded came to be known as the *Copenhagen interpretation*. The key idea was that a unique role was being played by classical measuring apparatus. Bohr held that it was the intervention of these large measuring devices that produced the determinating effect.

Even before the measurement issue surfaced, it had been necessary to have some way of seeing how one might recover from quantum theory the very considerable successes of classical mechanics in describing processes taking place on an everyday scale. It would be no use describing the microscopic at the expense of losing an understanding of the macroscopic. This requirement, called the *correspondence principle*, roughly amounted to being able to see that 'large' systems (the scale of largeness being set by Planck's constant) should behave in a way excellently approximated to by Newton's equations. Later, people came to realize that the relation between quantum mechanics and classical mechanics was a good deal subtler than this simple picture conveyed. Subsequently we shall see that there are some macroscopic phenomena that display certain intrinsically quantum properties, even including the possibility of technological exploitation, as in quantum computing. However, these arise in somewhat exceptional circumstances and the general drift of the correspondence principle was in the right direction.

Bohr emphasized that a measurement involved both the quantum entity and the classical measuring apparatus and he insisted that one should think of the mutual engagement of the two as being a single package deal (which he called a 'phenomenon'). Exactly where along the chain of correlated consequences leading from one end to the other the particularity of a specific result set in was then a question that could be avoided, as long as one kept the two ends of the chain inseparably connected together.

At first sight, there is something attractive about this proposal. If you go into a physics laboratory, you will find it littered with the kinds of apparatus of which Bohr spoke. Yet there is also something fishy about the proposal. Its account is dualistic in tone, as if the population of the physical world were made up of two different classes of being: fitful quantum entities and determinating classical measuring apparatus. In actual fact, however, there is a single, monistic physical world. Those bits of classical apparatus are themselves composed of quantum constituents (ultimately quarks, gluons, and electrons). The original Copenhagen interpretation failed to address the problem of how determinating apparatus could emerge from an indeterminate quantum substrate.

Nevertheless, it may be that Bohr and his friends were waving their hands in the right direction, even if not yet vigorously enough. Today, I think the majority of practising quantum physicists would subscribe to what one might call a neo-Copenhagen interpretation. On this view, the largeness and complexity of macroscopic apparatus is what somehow enables it to play the determinating role. How this happens is certainly not at all adequately understood, but at least one can correlate it with another (also not fully understood) property of large systems. This is their *irreversibility*.

With one exception that genuinely is not significant for the present discussion, the fundamental laws of physics are reversible. To see what this means, suppose, contrary to Heisenberg, that one could make a film of two electrons interacting. That film would make equal sense if it were run forwards or backwards. In other words, in the microworld, there is no intrinsic arrow of time, distinguishing the future from the past. In the macroworld, obviously, things are very different. Systems run down and the everyday world is irreversible. A film of a bouncing ball in which the bounces get higher and higher is being run backwards. These effects are related to the second law of thermodynamics, which states that, in an isolated system, entropy (the measure of disorder) never decreases. The reason this happens is that there are so many more ways of being disorderly than there are of being orderly, so that disarray wins hands down. Just think of your desk, if you do not intervene from time to time to tidy it up.

Now, measurement is the irreversible registration of a macroscopic signal of the state of affairs in the microworld. Therefore it incorporates an intrinsic direction of time: before there was no result, afterwards there is one. Thus there is some plausibility in supposing that an adequate understanding of large and complex systems that fully explained their irreversibility might also afford a valuable clue to the nature of the role they play in quantum measurement. In the current state of knowledge, however, this remains a pious aspiration rather than an actual achievement.

(3) New physics

Some have considered that solving the measurement problem will call for more radical thinking than simply pushing further principles that are already familiar to science. Ghirardi, Rimmer, and Weber have made a particularly interesting suggestion along these innovative lines (which has come to be known as the GRW theory). They propose that there is a universal property of random wavefunction collapse in space, but that the rate at which this happens depends on the amount of matter present. For quantum entities on their own, this rate is too tiny to have any noticeable effect, but in the presence of macroscopic quantities of matter (for instance, in a piece of classical measuring apparatus) it becomes so rapid as to be practically instantaneous.

This is a suggestion that, in principle, would be open to investigation through delicate experiments aimed at detecting other manifestations of this propensity to collapse. In the absence of such empirical confirmation, however, most physicists regard the GRW theory as too *ad hoc* to be persuasive.

(4) Consciousness

In the analysis of the Stern-Gerlach experiment, the last link in the correlated chain was a human observer hearing the counter click. Every quantum measurement of whose outcome we have actual knowledge has had as its final step someone's conscious awareness of the result. Consciousness is the ill-understood but undeniable (except by certain philosophers) experience of the interface between the material and the mental. The effects of drugs or brain damage make it clear that the material can act upon the mental. Why should we not expect a reciprocal power of the mental to act upon the material? Something like this seems to happen when we execute the willed intention of raising an arm. Perhaps, then, it is the intervention of a conscious observer that determines the outcome of a measurement. At first sight, the proposal has some attraction, and a number of very distinguished physicists have espoused this point of view. Nevertheless, it also has some very severe difficulties.

At most times and in most places, the universe has been devoid of consciousness. Are we to suppose that throughout these vast tracts of cosmic space and time, no quantum process resulted in a determinate consequence? Suppose one were to set up a computerized experiment in which the result was printed out on a piece of paper, which was then automatically stored away without any observer looking at it until six months later. Would it be the case that only at that subsequent time would there come to be a definite imprint on the paper?

These conclusions are not absolutely impossible, but many scientists do not find them at all plausible. The difficulties intensify further if we consider the sad story of Schrödinger's cat. The unfortunate animal is immured in a box that also contains a radioactive source with a 50–50 chance of decaying within the next hour. If the decay takes place, the emitted radiation will trigger the release of poison gas that will instantly kill the cat. Applying the conventional principles of quantum theory to the box and its contents leads to the implication that at the end of the hour, before a conscious observer lifts the lid of the box, the cat is in an even-handed superposition of 'alive' and 'dead'. Only after the box is opened will there be a collapse of possibilities, resulting in the discovery of either a definitely cooling corpse or a definitely frisking feline. But surely the animal knows whether or not it is alive, without requiring human intervention to help it to that conclusion? Perhaps we should conclude, therefore, that cat consciousness is as effective at determinating quantum outcomes as is human consciousness. Where then do we stop? Can worms also collapse the wavefunction? They may not exactly be conscious, but one would tend to suppose that in some way or another they have the definite property of being either alive or dead. These kinds of difficulties have prevented most physicists from believing that hypothesizing a unique role for consciousness is the way to solve the measurement problem.

(5) Many worlds

A yet more daring proposal rejects the idea of collapse altogether. Its proponents assert that the quantum formalism should be taken with greater seriousness than to impose upon it from outside the entirely *ad hoc* hypothesis of discontinuous change in the wavefunction. Instead one should acknowledge that everything that can happen *does* happen.

Why then do experimenters have the contrary impression, finding that on this occasion the electron is 'here' and nowhere else? The answer given is that this is the narrowly parochial view of an observer in this universe, but quantum reality is much greater than so constrained a picture suggests. Not only is there a world in which Schrödinger's cat lives, but there is also a parallel but disconnected world in which Schrödinger's cat dies. In other words, at every act of measurement, physical reality divides into a multiplicity of separate universes, in each of which different (cloned) experimenters observe the different possible outcomes of the measurement. Reality is a multiverse rather than a simple universe.

Since quantum measurements are happening all the time, this is a proposal of astonishing ontological prodigality. Poor William of Occam (whose logical 'razor' is supposed to cut out unnecessarily prodigal assumptions) must be turning in his grave at the thought of such a multiplication of entities. A different way of conceiving of this unimaginably immense proliferation is to locate it as happening not externally to the cosmos but internally to the mind/brain states of observers. Making that move is a turn from a many-worlds interpretation to a many-minds interpretation, but this scarcely serves to mitigate the prodigality of the proposal.

At first, the only physicists attracted to this way of thinking were the quantum cosmologists, seeking to apply quantum theory to the universe itself. While we remain perplexed about how the microscopic and the macroscopic relate to each other, this extension in the direction of the cosmic is a bold move whose feasibility is not necessarily obvious. If it is to be made, however, the many-worlds approach may seem the only option to use, for when the cosmos is involved there is no room left over for scientific appeal to the effects of external large systems or of consciousness. Latterly, there seems to have been a degree of widening inclination among other physicists to embrace the many-worlds approach, but for many of us it still remains a metaphysical steam hammer brought in to crack an admittedly tough quantum nut.

(6) Determinism

In 1954, David Bohm published an account of quantum theory that was fully deterministic, but which gave exactly the same experimental predictions as those of conventional quantum mechanics. In this theory, probabilities arise simply from ignorance of certain details. This remarkable discovery led John Bell to re-examine von Neumann's argument stating that this was impossible and to exhibit the flawed assumption on which this erroneous conclusion had been based.

Bohm achieved this impressive feat by divorcing wave and particle, which Copenhagen thinking had wedded in indissoluble complementarity. In the Bohm theory there are particles that are as unproblematically classical as even Isaac Newton himself would have wished them to be. When their positions or momenta are measured, it is simply a matter of observing what is unambiguously the case. In addition to the particles, however, there is a completely separate wave, whose form at any instant encapsulates information about the whole environment. This wave is not directly discernible but it has empirical consequences, for it influences the motion of the particles in a way that is additional to the effects of the conventional forces that may also act on them. It is this influence of the hidden wave (sometimes referred to as the 'guiding wave' or the source of the 'quantum potential') that sensitively affects the particles and succeeds in producing the appearance of interference effects and the characteristic probabilities associated with them. These guiding wave effects are strictly deterministic. Although the consequences are tightly predictable, they depend very delicately on the fine detail of the actual positions of the particles, and it is this sensitivity to minute variations that produces the appearance of randomness. Thus it is the particle positions that act as the hidden variables in Bohmian theory.

To understand Bohm's theory further, it is instructive to enquire how it deals with the double slits experiment. Because of the picturable nature of the particles, in this theory the electron must definitely go through one of the slits. What, then, was wrong with our earlier argument that this could not be so? What makes it possible to circumvent that earlier conclusion is the effect of the hidden wave. Without its independent existence and influence, it would indeed be true that if the electron went through slit A, slit B was irrelevant and could have been either open or shut. But Bohm's wave encapsulates instantaneous information about the total environment, and so its form is different if B is shut to what it is when B is open. This difference produces important consequences for the way in which the wave guides the particles. If B is shut, most of them are directed to the spot opposite A; if B is open, most

of them are directed towards the midpoint of the detector
screen.

One might have supposed that a determinate and picturable version
of quantum theory would have a great attraction for physicists. In
actual fact, few of them have taken to Bohmian ideas. The theory is
certainly instructive and clever, but many feel that it is too clever by
half. There is an air of contrivance about it that makes it
unappealing. For example, the hidden wave has to satisfy a wave
equation. Where does this equation come from? The frank answer
is out of the air or, more accurately, out of the mind of Schrödinger.
To get the right results, Bohm's wave equation must be the
Schrödinger equation, but this does not follow from any internal
logic of the theory and it is simply an *ad hoc* strategy designed to
produce empirically acceptable answers.

There are also certain technical difficulties that make the theory
seem less than totally satisfactory. One of the most challenging of
these relates to probabilistic properties. I have to admit that, for
simplicity, I have so far not quite stated these correctly. What is
exactly true is that if the *initial* probabilities relating to the particle
dispositions coincide with those that conventional quantum theory
would prescribe, then that coincidence between the two theories
will be maintained for all the subsequent motion. However, you
must start off right. In other words, the empirical success of Bohm's
theory requires either that the universe happened to start up with
the right (quantum) probabilities built in initially or, if it did not,
then some process of convergence quickly drove it in that direction.
This latter possibility is not inconceivable (a physicist would call it
'relaxation' onto the quantum probabilities), but it has not been
demonstrated nor has its timescale been reliably estimated.

The measurement problem continues to cause us anxiety as we
contemplate the bewildering range of, at best, only partially
persuasive proposals that have been made for its solution. Options
resorted to have included disregard (irrelevance); known physics

(decoherence); hoped-for physics (large systems); unknown new physics (GRW); hidden new physics (Bohm); metaphysical conjecture (consciousness; many worlds). It is a tangled tale and one that it is embarrassing for a physicist to tell, given the central role that measurement has in physical thinking. To be frank, we do not have as tight an intellectual grasp of quantum theory as we would like to have. We can do the sums and, in that sense, explain the phenomena, but we do not really *understand* what is going on. For Bohr, quantum mechanics is indeterminate; for Bohm, quantum mechanics is determinate. For Bohr, Heisenberg's uncertainty principle is an ontological principle of indeterminacy; for Bohm, Heisenberg's uncertainty principle is an epistemological principle of ignorance. We shall return to some of these metaphysical and interpretative questions in the final chapter. Meanwhile, a further speculative question awaits us.

Are there preferred states?

In the 19th century, mathematicians such as Sir William Rowan Hamilton developed very general understandings of the nature of Newtonian dynamical systems. A feature of the results of these researches was to establish that there are a great many equivalent ways in which the discussion might be formulated. It is often convenient for purposes of physical thinking to give a preferred role to picturing processes explicitly as occurring in space, but this is by no means a fundamental necessity. When Dirac developed the general principles of quantum theory, this democratic equality between different points of view was maintained in the new dynamics that resulted. All observables, and their corresponding eigenstates, had equal status as far as fundamental theory was concerned. The physicists express this conviction by saying that there is no 'preferred basis' (a special set of states, corresponding to a special set of observables, that are of unique significance).

Wrestling with the measurement problem has raised in the minds of some the question whether this no-preference principle should be maintained. Among the variety of proposals on the table, there is the feature that most of them seem to assign a special role to certain states, either as the end states of collapse or as the states which afford the perspectival illusion of collapse: in a (neo-)Copenhagen discussion centring on measuring apparatus, spatial position appears to play a special role as one speaks of pointers on scales or marks on photographic plates; similarly in the many-worlds interpretation, it is these same states that are the basis of division between the parallel worlds; in the consciousness interpretation, it is presumably the brain states that correspond to these perceptions that are the preferred basis of the matter/mind interface; the GRW proposal postulates collapse onto states of spatial position; Bohm's theory assigns a special role to the particle positions, minute details of which are the effective hidden variables of the theory. We should also note that decoherence is a phenomenon that occurs in space. If these are in fact indications of the need to revise previous democratic thinking, quantum mechanics would prove to have yet further revisionary influence to bring to bear on physics.

Chapter 4
Further developments

The hectic period of fundamental quantum discovery in the mid-1920s was followed by a long developmental period in which the implications of the new theory were explored and exploited. We must now take note of some of the insights provided by these further developments.

Tunnelling

Uncertainty relations of the Heisenberg type do not only apply to positions and momenta. They also apply to time and energy. Although energy is, broadly speaking, a conserved quantity in quantum theory – just as it is in classical theory – this is only so up to the point of the relevant uncertainty. In other words, there is the possibility in quantum mechanics of 'borrowing' some extra energy, provided it is paid back with appropriate promptness. This somewhat picturesque form of argument (which can be made more precise, and more convincing, by detailed calculations) enables some things to happen quantum mechanically that would be energetically forbidden in classical physics. The earliest example of a process of this kind to be recognized related to the possibility of tunnelling through a potential barrier.

The prototypical situation is sketched in figure 7, where the square 'hill' represents a region, entry into which requires the

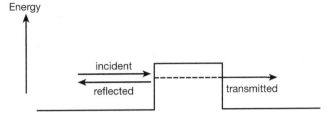

7. Tunnelling

payment of an energy tariff (called potential energy) equal to the
height of the hill. A moving particle will carry with it the energy of
its motion, which the physicists call kinetic energy. In classical
physics the situation is clear-cut. A particle whose kinetic energy is
greater than the potential energy tariff will sail across, traversing
the barrier at appropriately reduced speed (just as a car slows down
as it surmounts a hill), but then speeding up again on the other side
as its full kinetic energy is restored. If the kinetic energy is less than
the potential barrier, the particle cannot get across the 'hill' and it
must simply bounce back.

Quantum mechanically, the situation is different because of the
peculiar possibility of borrowing energy against time. This can
enable a particle whose kinetic energy is classically insufficient to
surmount the hill nevertheless sometimes to get across the barrier
provided it reaches the other side quickly enough to pay back
energy within the necessary time limit. It is as if the particle had
tunnelled through the hill. Replacing such picturesque story-telling
by precise calculations leads to the conclusion that a particle whose
kinetic energy is not too far below the height of the barrier will have
a certain probability of getting across and a certain probability of
bouncing back.

There are radioactive nuclei that behave as if they contain certain
constituents, called a-particles, that are trapped within the nucleus
by a potential barrier generated by the nuclear forces. These

particles, if they could only get through this barrier, would have enough energy to escape altogether on the other side. Nuclei of this type do, in fact, exhibit the phenomenon of a-decay and it was an early triumph of the application of quantum ideas at the nuclear level to use tunnelling calculations to give a quantitative account of the properties of such a-emissions.

Statistics

In classical physics, identical particles (two of a kind, such as two electrons) are nevertheless distinguishable from each other. If initially we label them 1 and 2, these marks of discrimination will have an abiding significance as we track along the separate particle trajectories. If the electrons eventually emerge after a complicated series of interactions, we can still, in principle, say which is 1 and which is 2. In the fuzzily unpicturable quantum world, in contrast, this is no longer the case. Because there are no continuously observable trajectories, all we can say after interaction is that *an* electron emerged here and *an* electron emerged there. Any initially chosen labelling cannot be followed through. In quantum theory, identical particles are also indistinguishable particles.

Since labels can have no intrinsic significance, the particular order in which they appear in the wavefunction (ψ) must be irrelevant. For identical particles, the (1,2) state must be physically the same as the (2,1) state. This does not mean that the wavefunction is strictly unchanged by the interchange, for it turns out that the same physical results would be obtained either from ψ or from $-\psi$ [**11**]. This little argument leads to a big conclusion. The result concerns what is called 'statistics', the behaviour of collections of identical particles. Quantum mechanically there are two possibilities (corresponding to the two possible signs of the behaviour of ψ under interchange):

bose statistics, holding in the case that ψ is unchanged under interchange. That is to say, the wavefunction is symmetric under exchange of two particles. Particles that have this property are called bosons.

fermi statistics, holding in the case that ψ changes sign under interchange. That is to say, the wavefunction is antisymmetric under exchange of two particles. Particles that have this property are called fermions.

Both options give behaviours that are different from the statistics of classically distinguishable particles. It turns out that quantum statistics leads to consequences that are of importance both for a fundamental understanding of the properties of matter and also for the technological construction of novel devices. (It is said that 30% of the United States GDP is derived from quantum-based industries: semiconductors, lasers, etc.)

Electrons are fermions. This implies that two of them can never be found in exactly the same state. This fact follows from arguing that interchange would produce both no change (since the two states are the same) and also a change of sign (because of fermi statistics). The only way out of this dilemma is to conclude that the two-particle wavefunction is actually zero. (Another way of stating the same argument is to point out that you cannot make an antisymmetric combination of two identical entities.) This result is called the *exclusion principle* and it provides the basis for understanding the chemical periodic table, with its recurring properties of related elements. In fact, the exclusion principle lies at the basis of the possibility of a chemistry sufficiently complex ultimately to sustain the development of life itself.

The chemical story goes like this: in an atom there are only certain energy states available for electrons and, of course, the exclusion principle requires that there should be no more than one electron occupying any one of them. The stable lowest energy state of the atom corresponds to filling up the least energetic states available. These states may be what the physicists call 'degenerate', meaning that there are several different states, all of which happen to have the same energy. A set of degenerate states constitutes an energy level. We can mentally picture the lowest energy state of the atom as being made up by adding electrons one by one to successive energy levels, up to the required number of electrons in the atom. Once all the states of a particular energy level are full, a further electron will have to go into the next highest energy level possessed by the atom. If that level in turn gets filled up, then on to the next level, and so on. In an atom with many electrons, the lowest energy levels (they are also called 'shells'), will all be full, with any electrons left over partially occupying the next shell. These 'left-over' electrons are the ones farthest from the nucleus and because of this they will determine the chemical interactions of the atom with other atoms. As one moves up the scale of atomic complexity (traversing the periodic table), the number of left-over electrons (0, 1, 2, . . .) varies cyclically, as shell after shell gets filled, and it is this repeating pattern of outermost electrons that produces the chemical repetitions of the periodic table.

In contrast to electrons, photons are bosons. It turns out that the behaviour of bosons is the exact opposite of the behaviour of fermions. No exclusion principle for them! Bosons like to be in the same state. They are similar to Southern Europeans, cheerfully crowding together in the same railway carriage, while fermions are like Northern Europeans, spread singly throughout the whole train. This mateyness of bosons is a phenomenon that in its most extreme form leads to a degree of concentration in a single state that is called bose condensation. It is this property that lies behind the technological device of the laser. The power of laser light is due to its being what is called 'coherent', that is to say, the light consists of many photons that are all in precisely the same state, a property that bose statistics strongly encourages. There are also effects associated with superconductivity (the vanishing of electrical resistance at very low temperatures) that depend on bose condensation, leading to macroscopically observable consequences of quantum properties. (The low temperature is required to prevent thermal jostling destroying coherence.)

Electrons and photons are also particles with spin. That is to say, they carry an intrinsic amount of angular momentum (a measure of rotational effects), almost as if they were little spinning tops. In the units that are natural to quantum theory (defined by Planck's constant), the electron has spin 1/2 and the photon has spin 1. It turns out that this fact exemplifies a general rule: particles of integer spin (0, 1, . . .) are always bosons; particles of half-odd integer spin (1/2, 3/2, . . .) are always fermions. From the point of view of ordinary quantum theory, this *spin and statistics theorem* is just an unexplained rule of thumb. However, it was discovered by Wolfgang Pauli (who also formulated the exclusion principle) that when quantum theory and special relativity are combined, the theorem emerges as a necessary consequence of that combination. Putting the two theories together yields richer insight than either provides on its own. The whole proves to be more than the sum of its parts.

Band structure

The form of solid matter that is simplest to think about is a crystal, in which the constituent atoms are ordered in the pattern of a regular array. A macroscopic crystal, significant on the scale of everyday experience, will contain so many atoms that it can be treated as effectively infinitely big from the microscopic point of view of quantum theory. Applying quantum mechanical principles to systems of this kind reveals new properties, intermediate between those of individual atoms and those of freely moving particles. We have seen that in an atom, possible electron energies come in a discrete series of distinct levels. On the other hand, a freely moving electron can have any positive energy whatsoever, corresponding to the kinetic energy of its actual motion. The energetic properties of electrons in crystals are a kind of compromise between these two extremes. The possible values of

8. **Band structure**

energy are found to lie within a series of bands. Within a band, there is a continuous range of possibilities; between bands, no energy levels at all are available to the electrons. In summary, the energetic properties of electrons in a crystal correspond to a series of alternating allowed and forbidden ranges of values.

The existence of this band structure provides the basis for understanding the electrical properties of crystalline solids. Electric currents result from inducing the movement of electrons within the solid. If the highest energy band of a crystal is totally full, this change of electron state will require exciting electrons across the gap into the band above. The transition would demand a significant energy input per excited electron. Since this is very hard to effect, a crystal with totally filled bands will behave as an insulator. It will be very difficult to induce motion in its electrons. If, however, a crystal has its highest band only partially filled, excitation will be easy, for it will only require a small energy input to move an electron into an available state of slightly higher energy. Such a crystal will behave as an electrical conductor.

Delayed choice experiments

Additional insight into the strange implications of the superposition principle was provided by John Archibald Wheeler's discussion of what he called 'delayed choice experiments'. A possible arrangement is shown in figure 9. A narrow beam of light is

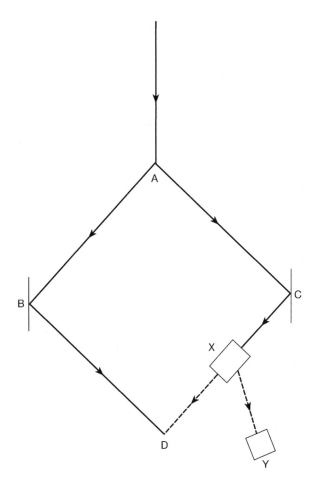

9. A delayed choice experiment

split at A into two sub-beams which are reflected by the mirrors at B and C to bring them together again at D, where an interference pattern can form due to the phase difference between the two paths (the waves have got out of step). One may consider an initial beam so weak that at any one time only a single photon is traversing the apparatus. The interference effects at D are then to be understood as due to self-interference between the two superposed states: (left-hand path) and (right-hand path). (Compare with the discussion of the double slit experiment in Chapter 2.) The new feature that Wheeler discussed arises if the apparatus is modified by inserting a device X between C and D. X is a switch that either lets a photon through or diverts it into a detector Y. If the switch is set for transmission, the experiment is the same as before, with an interference pattern at D. If the switch is set for deflection and the detector Y registers a photon, then there can be no interference pattern at D because that photon must definitely have taken the right-hand path for it to be deflected by Y. Wheeler pointed out the strange fact that the setting of X could be chosen while the photon is in flight *after* A. Until the switch setting is selected, the photon is, in some sense, backing two options: that of following both left- and right-hand paths and also that of following only one of them. Clever experiments have actually been conducted along these lines.

Sums over histories

Richard Feynman discovered an idiosyncratic way of reformulating quantum theory. This reformulation yields the same predictions as the conventional approach but offers a very different pictorial way of thinking about how these results arise.

Classical physics presents us with clear trajectories, unique paths of motion connecting the starting point A to the end point B. Conventionally these are calculated by solving the celebrated equations of Newtonian mechanics. In the 18th century, it was discovered that the actual path followed could be prescribed in a different, but equivalent, way by describing it as that trajectory joining A to B that gave the minimum value for a particular dynamical quantity associated with different paths. This quantity is called 'action' and its precise definition need not concern us here. The principle of least action (as it naturally came to be known) is akin to the property of light rays, that they take the path of minimal time between two points. (If there is no refraction, that path is a straight line, but in a refracting medium the least time principle leads to the familiar bending of the rays, as when a stick in a glass of water appears bent.)

Because of the cloudy unpicturability of quantum processes, quantum particles do not have definite trajectories. Feynman suggested that, instead, one should picture a quantum particle as moving from A to B *along all possible paths*, direct or wriggly, fast or slow. From this point of view, the wavefunction of conventional thinking arose from adding together contributions from all these possibilities, giving rise to the description of 'sums over histories'.

The details of how the terms in this immense sum are to be formed are too technical to be gone into here. It turns out that the contribution from a given path is related to the action associated with that path, divided by Planck's constant. (The physical dimensions of action and of h are the same, so their ratio is a pure number, independent of the units in which we choose to measure physical quantities.) The actual form taken by these contributions from different paths is such that neighbouring paths tend to cancel each other out, due to rapid fluctuations in the signs (more accurately, phases) of their contributions. If the system being considered is one whose action is large with respect to h, only the path of minimal action will contribute much (since it turns out that it is near that path that the fluctuations are the least and so the effect of cancellations is minimized). This observation provides a simple way of understanding why large systems behave classically, following paths of least action.

Formulating these ideas in a precise and calculable way is not at all easy. One may readily imagine that the range of variation represented by the multiplicity of possible paths is not a simple aggregate over which to sum. Nevertheless, the sums over histories approach has had two important consequences. One is that it led Feynman on to discover a much more manageable calculational technique, now universally called 'Feynman integrals', which is the most useful approach to quantum calculations made available to physicists in the last 50 years. It yields a physical picture in which interactions are due to the exchange of energy and momentum carried across by what are called *virtual particles*. The adjective 'virtual' is used because these intermediate 'particles', which cannot appear in the initial or final states of the process, are not constrained to have physical masses, but rather one sums over all possible mass values.

The other advantage of the sums over histories approach is that there are some rather subtle and tricky quantum systems for which it offers a clearer way of formulating the problem than that given by the more conventional approach.

More about decoherence

The environmental effects of ubiquitous radiation that produce decoherence have a significance that goes beyond their partial relevance to the measurement problem. One important recent development has been the realization that they also bear on how one should think about the quantum mechanics of so-called chaotic systems.

The intrinsic unpredictabilities that are present in nature do not arise solely from quantum processes. It was a great surprise to most physicists when, some 40 years ago, the realization dawned that even in Newtonian physics there are many systems whose extreme sensitivity to the effects of very small disturbances makes their future behaviour beyond our power to predict accurately. These chaotic systems (as they are called) soon come to be sensitive to detail at the level of Heisenberg uncertainty or below. Yet their treatment from a quantum point of view – a subject called *quantum chaology* – proves to be problematic.

The reason for perplexity is as follows: chaotic systems have a behaviour whose geometrical character corresponds to the celebrated fractals (of which the Mandelbrot set, the subject of a hundred psychedelic posters, is the best-known example). Fractals are what is called 'self-similar', that is to say, they look essentially the same on whatever scale they are examined (saw-teeth made out of saw-teeth, . . . , all the way down). Fractals, therefore, have no natural scale. Quantum systems, on the other hand, do have a natural scale, set by Planck's constant. Therefore chaos theory and quantum theory do not fit smoothly onto each other.

The resulting mismatch leads to what is called 'the quantum suppression of chaos': chaotic systems have their behaviour modified when it comes to depend on detail at the quantum level. This in turn leads to another problem for the physicists, arising in its most acute form from considering the 16th moon of Saturn, called Hyperion. This potato-shaped lump of rock, about the size of New York, is tumbling about in a chaotic fashion. If we apply notions of quantum suppression to Hyperion, the result is expected to be astonishingly effective, despite the moon's considerable size. In fact, on the basis of this calculation, chaotic motion could only last at most for about 37 years. In actual fact, astronomers have been observing Hyperion for rather less time than that, but no one expects that its weird tumbling is going to come to an end quite soon. At first sight, we are faced with a serious problem. However, taking decoherence into account solves it for us. Decoherence's tendency to move things in a more classical-seeming direction has the effect, in its turn, of suppressing the quantum suppression of chaos. We can confidently expect Hyperion to continue tumbling for a very long time yet.

Another effect of a rather similar kind due to decoherence is the *quantum Zeno effect*. A radioactive nucleus due to decay will be forced back to its initial state by the 'mini-observations' that result from its interaction with environmental photons. This continual return to square one has the effect of inhibiting the decay, a phenomenon that has been observed experimentally. This effect is named after the ancient Greek philosopher Zeno, whose meditation on observing an arrow to be *now* at a particular fixed point persuaded him that the arrow could not really be moving.

These phenomena make it clear that the relationship between quantum theory and its classical limit is subtle, involving the interlacing of effects that cannot be characterized just by a simplistic division into 'large' and 'small'.

Relativistic quantum theory

Our discussion of the spin and statistics theorem has already shown that putting quantum theory and special relativity together produces a unified theory of enriched content. The first successful equation that succeeded in consistently formulating the combination of the two was the relativistic equation of the electron, discovered by Paul Dirac in 1928 [**12**]. Its mathematical detail is too technical to be presented in a book of this kind, but we must note two important and unanticipated consequences that flowed from this development.

Dirac produced his equation simply with the needs of quantum theory and of relativistic invariance in mind. It must have been a gratifying surprise, therefore, when he found that the equation's predictions of the electromagnetic properties of the electron were such that it turned out that the electron's magnetic interactions were twice as strong as one would have naively expected them to be on the basis of thinking of the electron as being a miniature, electrically charged, spinning top. It was already known empirically that this was the case, but no one had been able to understand why this apparently anomalous behaviour should be so.

The second, even more significant, consequence resulted from Dirac's brilliantly turning threatened defeat into triumphant victory. As it stood, the equation had a gross defect. It allowed positive energy states of the kind one needed to correspond to the behaviour of actual electrons, but it allowed negative energy states as well. The latter just did not make physical sense. Yet they could not simply be discarded, for the principles of quantum mechanics would inevitably allow the disastrous consequence of transitions to them from the physically acceptable positive energy states. (This would be a physical disaster because transitions to such states could produce unlimited quantities of counterbalancing positive energy, resulting in a kind of runaway perpetual motion machine.) For quite a while this was a highly embarrassing puzzle. Then Dirac realized that the fermi statistics of electrons might permit a way out of the dilemma. With great boldness, he supposed that all the negative energy states were already occupied. The exclusion principle would then block the possibility of any transitions to them from the positive energy states. What people had thought of as empty space (the vacuum) was in fact filled with this 'sea' of negative-energy electrons!

It sounds rather an odd picture and later, in fact, it proved possible to formulate the theory in a way that preserved the desirable results in a manner less picturesque but also less weird. In the meantime, working with the concept of the negative-energy sea led Dirac to a discovery of prime importance. If enough energy were to be provided, say by a very energetic photon, it would be possible to eject a negative-energy electron from the sea, turning it into a positive-energy electron of the ordinary kind. What then was one to make of the 'hole' that this process had left behind in the negative sea? The absence of negative energy is the same as the presence of positive energy (two minuses make a plus), so the hole would behave like a positive-energy particle. But the absence of negative charge is the same as the presence of positive charge, so this 'hole-particle' would be positively charged, in contrast to the negatively charged electron.

In the 1930s, the thinking of elementary particle physicists was pretty conservative compared to the speculative freedom that was to come later. They did not at all like the idea of suggesting the existence of some new, hitherto unknown type of particle. Initially, therefore, it was supposed that this positive particle that Dirac had come to talk about might simply be the well-known, positively charged proton. However, it was soon realized that the hole had to have the same mass as the electron, while the proton is very much more massive. Thus the only acceptable interpretation on offer led to the somewhat reluctant prediction of a totally new particle, quickly christened the positron, of electronic mass but positive charge. Soon its existence was experimentally confirmed as positrons were detected in cosmic rays. (In fact, examples had been seen much earlier, but they had not been recognized as such. Experimenters find it hard to see what they are not actually looking for.)

It came to be realized that this electron-positron twinning was a particular example of behaviour widespread in nature. There is both matter (such as electrons) and oppositely charged *antimatter* (such as positrons). The prefix 'anti' is appropriate because an electron and a positron can annihilate each other, disappearing in a burst of energy. (In the old-fashioned manner of speaking, the electron fills up the hole in the sea and the energy released is then radiated away. Conversely, as we have seen, a highly energetic photon can drive an electron out of the sea, leaving a hole behind and thereby creating an electron-positron pair.)

The fruitful history of the Dirac equation, leading both to an explanation of magnetic properties and to the discovery of antimatter, topics that played no part at all in the original motivation for the equation, is an outstanding example of the long-term value that can be displayed by a really fundamental scientific idea. It is this remarkable fertility that persuades physicists that they are really 'on to something' and that, contrary to the suggestions of some philosophers and sociologists of science, they are not just tacitly agreeing to look at things in a particular way. Rather, they are making discoveries about what the physical world is actually like.

Quantum field theory

Another fundamental discovery was made by Dirac when he applied the principles of quantum mechanics not to particles but to the electromagnetic field. This development yielded the first-known example of a quantum field theory. With hindsight, to make this step is not too difficult technically. The principal difference between a particle and a field is that the former has only a finite number of degrees of freedom (independent ways in which its state can change), while a field has an infinite number of degrees of freedom. There are well-known mathematical techniques for handling this difference.

Quantum field theories prove to be of considerable interest and afford us a most illuminating way of thinking about wave/particle duality. A field is an entity spread out in space and time. It is, therefore, an entity that has an intrinsically wavelike character. Applying quantum theory to the field results in its physical quantities (such as energy and momentum) becoming present in discrete, countable packets (quanta). But such countability is just what we associate with particlelike behaviour. In studying a quantum field, therefore, we are investigating and understanding an entity that explicitly exhibits both wavelike and particlelike properties in as clear a manner as possible. It is a bit like puzzling over how a mammal could come to lay an egg and then being shown a duck-billed platypus. An actual example is always maximally instructive. It turns out that in quantum field theory the states that show wavelike properties (technically, have definite phases) are those that contain an *indefinite* number of particles. This latter property is a natural possibility because of quantum theory's superposition principle that allows the combination of states with different numbers of particles in them. It would be an impossible option in classical theory, where one could just look and see in order to count the number of particles actually present.

The vacuum in quantum field theory has unusual properties that are particularly important. The vacuum is, of course, the lowest energy state, in which there will be no excitations present that correspond to particles. Yet, though in this sense there is nothing there, in quantum field theory it does not mean that there is nothing going on. The reason is as follows: a standard mathematical technique, called Fourier analysis, allows us to regard a field as equivalent to an infinite collection of harmonic oscillators. Each oscillator has a particular frequency associated with it and the oscillator behaves dynamically just as if it were a pendulum of that given frequency. The field vacuum is the state in which all these 'pendula' are in their lowest energy states. For a classical pendulum, that is when the bob is at rest and at the bottom. This is truly a situation in which nothing is happening. However, quantum mechanics does not permit so perfect a degree of tranquillity. Heisenberg will not allow the 'bob' to have both a definite position (at the bottom) and a definite momentum (at rest). Instead the quantum pendulum must be slightly in motion even in its lowest energy state (near the bottom and nearly at rest, but not quite). The resultant quantum quivering is called *zero-point motion*. Applying these ideas many times over to the infinite array of oscillators that is a quantum field implies that its vacuum is a humming hive of activity. Fluctuations continually take place, in the course of which transient 'particles' appear and disappear. A quantum vacuum is more like a plenum than like empty space.

When the physicists came to apply quantum field theory to situations involving interactions between fields, they ran into difficulties. The infinite number of degrees of freedom tended to produce infinite answers for what should have been finite physical quantities. One important way in which this happened was through interaction with the restlessly fluctuating vacuum. Eventually a way was found for producing sense out of nonsense. Certain kinds of field theories (called *renormalizable* theories) produce only limited kinds of infinities, simply associated with the masses of particles and with the strengths of their interactions. Just striking out these infinite terms and replacing them with the finite measured values of the relevant physical quantities is a procedure that defines meaningful results, even if the procedure is not exactly mathematically pure. It also turns out to provide finite expressions that are in stunning agreement with experiment. Most physicists are happy with this pragmatic success. Dirac himself was never so. He strongly disapproved of dubious legerdemain with formally infinite quantities.

Today, all theories of elementary particles (such as the quark theory of matter) are quantum field theories. Particles are thought of as energetic excitations of the underlying field. (An appropriate field theory also turns out to provide the right way to deal with the difficulties of the 'sea' of negative-energy electrons.)

Quantum computing

Recently there has been considerable interest in the possibility of exploiting the superposition principle as a means for gaining greatly enhanced computing power.

Conventional computing is based on the combination of binary operations, expressed formally in logical combinations of 0s and 1s, and realizable in hardware terms by switches that are either on or off. In a classical device, of course, the latter are mutually exclusive possibilities. A switch is either definitely on or definitely off. In the quantum world, however, the switch could be in a state that is a superposition of these two classical possibilities. A sequence of such superpositions would correspond to a wholly novel kind of parallel processing. The ability to keep so many computational balls in the air simultaneously could, in principle, represent an increase in computing power which the addition of extra elements would multiply up exponentially, compared with the linear increase in conventional circumstances. Many computational tasks, such as decodings or the factorization of very large numbers, would become feasible that are infeasible with present machines.

These are exciting possibilities. (Their proponents display a liking for talking about them in many-worlds terms, as if processing would take place in parallel universes, but it seems that really it is just the superposition principle itself that is the basis for the feasibility of quantum computing.) Actual implementation, however, will be a distinctly tricky business, with many problems yet to be solved. Many of these centre on the stable preservation of the superposed states. The phenomenon of decoherence shows how problematic it could be to insulate a quantum computer from deleterious environmental interference. Quantum computing is being given serious technological and entrepreneurial consideration, but as an effective procedure it currently remains a gleam in the eye of its proponents.

Chapter 5
Togetherness

Einstein, through his explanation of the photoelectric effect, had been one of the grandfathers of quantum theory. However, he came to detest his grandchild. Like the vast majority of physicists, Einstein was deeply convinced of the reality of the physical world and trusted in the truthful reliability of science's account of its nature. But he came to believe that this reality could only be guaranteed by the kind of naive objectivity that Newtonian thinking had assumed. In consequence, Einstein abhorred the cloudy fitfulness that Copenhagen orthodoxy assigned to the nature of the quantum world.

His first onslaught on modern quantum theory took the form of a series of highly ingenious thought experiments, each of which purported in some way to circumvent the limitations of the Heisenberg uncertainty principle. Einstein's opponent in this contest was Niels Bohr, who each time succeeded in showing that a thorough-going application of quantum ideas to all aspects of the proposed experiment actually resulted in the uncertainty principle surviving unscathed. Eventually Einstein conceded defeat in this particular battle.

After licking his wounds for a while, Einstein returned to the fray, staking out a new ground for contention. With two younger collaborators, Boris Podolsky and Nathan Rosen, he showed that there were some very peculiar, hitherto unnoticed, long-range implications for the quantum mechanical behaviour of two apparently well-separated particles. The issues are most easily explained in terms of a later development of what, bearing in mind its discoverers' names, we may call EPR thinking. The argument was due to David Bohm and, although it is a little involved, it is well worth wrestling with.

Suppose two particles have spins s_1 and s_2 and it is known that the total spin is zero. This implies, of course, that s_2 is $-s_1$. Spin is a vector (that is, it has magnitude and direction – think of it as an arrow) and we have followed mathematical convention in using boldface type for vectorial quantities. A spin vector will, therefore, have three components measured along three chosen spatial directions, x, y, and z. If one were to measure the x component of s_1 and get the answer s'_{1x}, then the x component of s_2 must be $-s'_{1x}$. If, on the other hand, one had measured the y component of s_1, getting the answer s'_{1y}, one would know that the y component of s_2 would have to be $-s'_{1y}$. But quantum mechanics does not permit one to measure both the x and y components of spin simultaneously, because there is an uncertainty relation between them. Einstein argued that, while this might be the case according to orthodox quantum thinking, whatever happened to particle 1 could have no immediate effect upon the distant particle 2. In EPR thinking, the spatial separation of 1 and 2 implies *the independence of what happens at 1 and what happens at 2*. If that is so, and if one can choose to measure either the x or the y components of spin at 1 and get certain knowledge of the x or y components respectively of spin at 2, then Einstein claimed that particle 2 must actually have these definite values for its spin components, whether the measurements were actually made or not. This was something that conventional quantum theory denied, because, of course, the uncertainty relation between x and y spin components applied as much to particle 2 as to particle 1.

Einstein's conclusion from this mildly complicated argument was that there must be something incomplete in conventional quantum theory. It failed to account for what he believed must be definite values of spin components. Almost all his fellow physicists interpret things differently. In their view, neither s_1 nor s_2 have definite spin components until a measurement is actually made. Then, determining the x component of 1 forces the x component of 2 to take the opposite value. That is to say, the measurement at 1 also forces a collapse of the wavefunction at 2 onto the opposite value of the x spin component. If it had been the y component that had been measured at 1, then the collapse at 2 would have been onto the opposite y spin component. These two 2-states (known x component; known y component) are absolutely distinct from each other. Thus the majority view leads to the conclusion that *measurement on 1 produces instantaneous change at 2, a change that depends precisely on exactly what is measured at 1*. In other words, there is some counterintuitive togetherness-in-separation between 1 and 2; action at 1 produces immediate consequences for 2 and the consequences are different for different actions at 1. This is usually called the *EPR effect*. The terminology is somewhat ironic since Einstein himself refused to believe in such a long-range connection, regarding it as an influence that was too 'spooky' to be acceptable to a physicist. There the matter rested for a while.

The next step was taken by John Bell. He analysed what properties the 1–2 system would have if it were a genuinely separated system (as Einstein had supposed), with properties at 1 depending only on what happened locally at 1 and properties at 2 depending only on what happened locally at 2. Bell showed that if this strict locality were the case, there would be certain relations between measurable quantities (they are now called the *Bell inequalities*) that quantum mechanics predicted would be violated in certain circumstances. This was a very significant step forward, moving the argument on from the realm of thought experiments into the empirically accessible realm of what could actually be investigated in the laboratory. The experiments were not easy to do but eventually, in the early 1980s, Alain Aspect and his collaborators were able to carry out a skilfully instrumented investigation that vindicated the predictions of quantum theory and negated the possibility of a purely local theory of the kind that Einstein had espoused. It had become clear that there is an irreducible degree of non-locality present in the physical world. Quantum entities that have interacted with each other remain mutually entangled, however far they may eventually separate spatially. It seems that nature fights back against a relentless reductionism. Even the subatomic world cannot be treated purely atomistically.

The EPR effect's implication of deep-seated relationality present in the fundamental structure of the physical world is a discovery that physical thinking and metaphysical reflection have still to come to terms with in fully elucidating all its consequences. As part of that continuing process of assimilation, it is necessary to be as clear as possible about what is the character of the entanglement that EPR implies. One must acknowledge that a true case of action at a distance is involved, and not merely some gain in additional knowledge. Putting it in learned language, the EPR effect is ontological and not simply epistemological. Increase in knowledge at a distance is in no way problematic or surprising. Suppose an urn contains two balls, one white, the other black. You and I both put in our hands and remove one of the balls in our closed fists. You then go a mile down the road, open your fist and find that you have the white ball. Immediately you know I must have the black one. The only thing that has changed in this episode is your state of knowledge. I always had the black ball, you always had the white ball, but now you have become aware that this is so. In the EPR effect, by contrast, what happens at 1 *changes* what is the case at 2. It is as if, were you to find that you had a red ball in your hand, I would have to have a blue ball in mine, but if you found a green ball, I would have to have a yellow ball and, previous to your looking, neither of us had balls of determinate colours.

An alert reader may query all this talk about instantaneous change. Does not special relativity prohibit something at 1 having any effect at 2 until there has been time for the transmission of an influence moving with at most the velocity of light? Not quite. What relativity actually prohibits is the instantaneous transmission of information, of a kind that would permit the immediate synchronization of a clock at 2 with a clock at 1. It turns out that the EPR kind of entanglement does not permit the conveyance of messages of that kind. The reason is that its togetherness-in-separation takes the form of correlations between what is happening at 1 and what is happening at 2 and no message can be read out of these correlations without knowledge of what is happening at both ends. It is as if a singer at 1 was singing a random series of notes and a singer at 2 was also singing a random series of notes and only if one were able to hear them both together would one realize that the two singers were in some kind of harmony with each other. Realizing this is so warns us against embracing the kind of 'quantum hype' argument that incorrectly asserts that EPR 'proves' that telepathy is possible.

Chapter 6
Lessons and meanings

The picture of physical process presented to us by quantum theory is radically different from what everyday experience would lead us to expect. Its peculiarity is such as to raise with some force the question of whether this is indeed what subatomic nature is 'really like' or whether quantum mechanics is no more than a convenient, if strange, manner of speaking that enables us to do the sums. We may get answers that agree startlingly well with the results obtained by the laboratory use of classical measuring apparatus, but perhaps we should not actually believe the theory. The issue raised is essentially a philosophical one, going beyond what can be settled simply by the use of science's own unaided resources. In fact, this quantum questioning is just a particular example – if an unusually challenging one – of the fundamental philosophical debate between the realists and the positivists.

Positivism and realism

Positivists see the role of science as being the reconciliation of observational data. If one can make predictions that accurately and harmoniously account for the behaviour of the measuring apparatus, the task is done. Ontological questions (What is really there?) are an irrelevant luxury and best discarded. The world of the positivist is populated by counter readings and marks on photographic plates.

This point of view has a long history. Cardinal Bellarmine urged upon Galileo that he should regard the Copernican system as simply a convenient means for 'saving the appearances', a good way of doing calculations to determine where planets would appear in the sky. Galileo should not think that the Earth actually went round the Sun – rather Copernicus should be considered as having used the supposition simply as a handy calculational device. This face-saving offer did not appeal to Galileo, nor have similar suggestions been favourably received by scientists generally. If science is just about correlating data, and is not telling us what the physical world is actually like, it is difficult to see that the enterprise is worth all the time and trouble and talent expended upon it. Its achievements would seem too meagre to justify such a degree of involvement. Moreover, the most natural explanation of a theory's ability to save appearances would surely be that it bore some correspondence to the way things are.

Nevertheless, Niels Bohr often seemed to speak of quantum theory in a positivistic kind of way. He once wrote to a friend that

> There is no quantum world. There is only abstract quantum physical description. It is wrong to think that the task of physics is to find out how nature *is*. Physics is concerned with what we can say about nature.

Bohr's preoccupation with the role of classical measuring apparatus could be seen as having encouraged such a positivistic-sounding point of view. We have seen that in his later years he became very concerned with philosophical issues, writing extensively about them. The resulting corpus is hard to interpret. Bohr's gift in philosophical matters fell far short of his outstanding talent as a physicist. Moreover, he believed – and exemplified – that there are two kinds of truth: a trivial kind, which could be articulated clearly, and a profound kind which could only be spoken about cloudily. Certainly the body of his writings has been very variously interpreted by the commentators. Some have felt that there was, in fact, a kind of qualified realism to which Bohr was an adherent.

Realists see the role of science to be to discover what the physical world is actually like. This is a task that will never be completely fulfilled. New physical regimes (encountered at yet higher energies, for example) will always be awaiting investigation, and they may well prove to have very unexpected features in their behaviour. An honest assessment of the achievement of physics can at most claim verisimilitude (an accurate account of a wide but circumscribed range of phenomena) and not absolute truth (a total account of physical reality). Physicists are the mapmakers of the physical world, finding theories that are adequate on a chosen scale but not capable of describing every aspect of what is going on. A philosophical view of this kind sees the attainment of physical science as being the tightening grasp of an actual reality. The world of the realist is populated by electrons and photons, quarks and gluons.

A kind of halfway house between positivism and realism is offered by pragmatism, the philosophical position that acknowledges the technological fact that physics enables us to get things done, but which does not go as far as a realist position in thinking that we know what the world is actually like. A pragmatist might say that we should take science seriously but we should not go so far as to believe it. Yet, far and away the most obvious explanation of the technological success of science is surely that it is based on a verisimilitudinous understanding of the way that matter actually behaves.

A number of defences of scientific realism can be mounted. One, already noted, is that it provides a natural understanding of the predictive successes of physics and its long-term fruitfulness, and of the reliable working of the many technological devices constructed in the light of its picture of the physical world. Realism also explains why scientific endeavour is seen to be worthwhile, attracting the lifelong devotion of many people of high talent, for it is an activity that yields actual knowledge of the way things are. Realism corresponds to the conviction of scientists that they experience the making of discoveries and that they are not just learning better ways to do the sums, or just tacitly agreeing among themselves to see things this way. This conviction of discovery arises powerfully from frequent experience of the recalcitrance displayed by nature in the face of the scientist's prior expectation. The physicist may approach phenomena with certain ideas in mind, only to find that they are negated by the actual way that the physical world is found to behave. Nature forces reconsideration upon us and this often drives the eventual discovery of the totally unexpected character of what is going on. The rise of quantum theory is, of course, an outstanding example of the revisionism imposed by physical reality upon the thinking of the scientist.

If quantum theory is indeed telling us what the subatomic world is really like, then its reality is something very different from the naive objectivity with which we can approach the world of everyday objects. This is the point that Einstein found so hard to accept. He passionately believed in the reality of the physical world but he rejected conventional quantum theory because he wrongly supposed that only the objective could be the real.

Quantum reality is cloudy and fitful in its character. The French philosopher-physicist Bernard d'Espagnat has spoken of its nature as being 'veiled'. The most truly philosophically reflective of the founding figures of quantum theory was Werner Heisenberg. He felt that it would be valuable to borrow from Aristotle the concept of *potentia*. Heisenberg once wrote that

> In experiments about atomic events we have to do with things that are facts, with phenomena that are just as real as any phenomena in daily life. But the atoms or elementary particles are not as real; they form a world of potentialities or possibilities rather than of things or facts.

An electron does not all the time possess a definite position or a definite momentum, but rather it possesses the potentiality for exhibiting one or other of these if a measurement turns the potentiality into an actuality. I would disagree with Heisenberg in thinking that this fact makes an electron 'not as real' as a table or a chair. The electron simply enjoys a different kind of reality, appropriate to its nature. If we are to know things as they are, we must be prepared to know them as they actually are, on their own terms, so to speak.

Why is it that almost all physicists want to insist on the reality, appropriately understood, of electrons? I believe it is because the assumption that there are electrons, with all the subtle quantum properties that go with them, makes intelligible great swathes of physical experience that otherwise would be opaque to us. It explains the conduction properties of metals, the chemical properties of atoms, our ability to build electron microscopes, and much else besides. It is *intelligibility* (rather than objectivity) that is the clue to reality – a conviction, incidentally, that is consonant with a metaphysical tradition stemming from the thought of Thomas Aquinas.

The veiled reality that is the essence of the nature of electrons is represented in our thinking by the wavefunctions associated with them. When a physicist thinks about what an electron is 'doing', it is the appropriate wavefunction that is in mind. Obviously the wavefunction is not as accessible an entity as the objective presence of a billiard ball, but neither does it function in quantum thinking in a way that makes one comfortable with the positivistic notion that it is simply a calculational device. The rather wraithlike wavefunction seems an appropriate vehicle to be the carrier of the veiled potentiality of quantum reality.

Reasonableness

If the study of quantum physics teaches one anything, it is that the world is full of surprises. No one would have supposed beforehand that there could be entities that sometimes behaved as if they were waves and sometimes behaved as if they were particles. This realization was forced upon the physics community by the intransigent necessity of actual empirical experience. As Bohr once said, the world is not only stranger than we thought; it is stranger than we could think. We noted earlier that even logic has to be modified when it is applied to the quantum world.

A slogan for the quantum physicist might well be 'No undue tyranny of commonsense'. This stirring motto conveys a message of wider relevance than to the quantum realm alone. It reminds us that our powers of rational prevision are pretty myopic. The instinctive question that a scientist ought to ask about a proposed account of some aspect of reality, whether within science or beyond it, is not 'Is it reasonable?', as if we felt we knew beforehand what form reason was bound to take. Rather, the proper question is 'What makes you think this might be the case?' The latter is a much more open question, not foreclosing the possibility of radical surprise but insisting that there should be evidential backing for what is being asserted.

If quantum theory encourages us to keep fluid our conception of what is reasonable, it also encourages us to recognize that there is no universal epistemology, no single sovereign way in which we may hope to gain all knowledge. Although we can know the everyday world in its Newtonian clarity, we can only know the quantum world if we are prepared to accept it in its Heisenbergian uncertainty. Insisting on a naively objective account of electrons can only lead to failure. There is a kind of epistemological circle: how we know an entity must conform to the nature of that entity; the nature of the entity is revealed through what we know about it. There can be no escape from this delicate circularity. The example of quantum theory encourages the belief that the circle can be benign and not vicious.

Metaphysical criteria

Successful physical theories must eventually be able to exhibit their ability to fit the experimental facts. The ultimate saving of appearances is a necessary achievement, though there may be some interim periods of difficulty on the way to that end (as when Dirac initially faced the apparently empirically disastrous prediction of negative energy states of the electron). Particularly persuasive will be the property of sustained fruitfulness, as a theory proves able to predict or give understanding of new or unexpected phenomena (Dirac's explanation of the magnetic properties of electrons and his prediction of the positron).

Yet these empirical successes are not by themselves always sufficient criteria for the endorsement of a theory by the scientific community. The choice between an indeterministic interpretation of quantum theory and a deterministic interpretation cannot be made on these grounds. Bohm saves appearances as well as Bohr does. The issue between them has to be settled for other reasons. The decision turns out to depend upon metaphysical judgement and not simply on physical measurements.

Metaphysical criteria that the scientific community take very seriously in assessing the weight to put on a theory include:

(1) Scope

The theory must make intelligible the widest possible range of phenomena. In the case of Bohr and Bohm, this criterion does not lead to a settlement of the issue between them, because of the empirical equivalence of the two sets of results (though one should note that Bohmian thinking needs to complete its account by better arguments to substantiate its belief that the initial probabilities are correctly given by a wavefunction calculation).

(2) Economy

The more concise and parsimonious a theory is, the more attractive it will seem. Bohm's theory scores less well here because of its assumption of the hidden wave in addition to the observable particles. This multiplication of entities is certainly seen by many physicists to be an unattractive feature of the theory.

(3) Elegance

This is a notion, to which one can add the property of *naturalness*, that results from the lack of undue contrivance. It is on these grounds that most physicists find the greatest difficulty with Bohmian ideas. In particular, the *ad hoc* but necessary appropriation of the Schrödinger equation as the equation for the Bohmian wave has an unattractively opportunist air to it.

These criteria do not only lie outside physics itself, but they are also such that their assessment is a matter of personal judgement. That they are satisfied is not a matter that can be reduced to following a formalized protocol. It is not a judgement whose evaluation could be delegated to a computer. The majority verdict of the quantum physics community in favour of Bohr and against Bohm is a paradigm example of what the philosopher of science, Michael Polanyi, would have referred to as the role in science of 'personal knowledge'. Polanyi, who had himself been a distinguished physical chemist before he turned to philosophy, emphasized that, though the subject matter of science is the impersonal physical world, the activity of doing science is ineluctably an activity of persons. This is because it involves many acts of judgement that require the exercise of tacit skills that can only be acquired by persons who have served a long apprenticeship within the truth-seeking community of scientists. These judgements do not only concern the application of the kind of metaphysical criteria we have been discussing. At a more everyday level they include such skills as the experimenter's ability to assess and eliminate spurious 'background' effects that might otherwise contaminate the results of an experiment. There is no little black book that tells the experimenter how to do this. It is something learned from experience. In a phrase that Polanyi often repeated, we all 'know more than we can tell', whether this is shown in the tacit skills of riding a bicycle, the connoisseurship of wine, or the design and execution of successful physical experiments.

Holism

We have seen in Chapter 5 that the EPR effect shows that there is an intrinsic non-locality present in the quantum world. We have also seen that the phenomenon of decoherence has made plain the quite astonishingly powerful effects that the general environment can exercise on quantum entities. Although quantum physics is the physics of the very small, it by no means endorses a purely atomistic, 'bits and pieces', account of reality.

Physics does not determine metaphysics (the wider world view), but it certainly constrains it, rather as the foundations of a house constrain, but do not determine completely, the edifice that will be built upon them. Philosophical thinking has not always adequately taken into account the implications of these holistic aspects of quantum theory. There can be no doubt that they encourage acceptance of the necessity of attaining an account of the natural world that succeeds both in recognizing that its building blocks are indeed elementary particles, but also that their combination gives rise to a more integrated reality than a simple constituent picture on its own would suggest.

The role of the observer

A cliché that is often repeated is that quantum theory is 'observer created'. More careful thought will considerably qualify and reduce that claim. What can be said will depend critically upon what interpretation of the measurement process is chosen. This is the central issue because, between measurements, the Schrödinger equation prescribes that a quantum system evolves in a perfectly continuous and determined fashion. It is also important to recall that the general definition of measurement is the irreversible macroscopic registration of the signal of a microscopic state of affairs. This happening may involve an observer, but in general it need not.

Only the consciousness interpretation assigns a unique role to the acts of a conscious observer. All other interpretations are concerned simply with aspects of physical process, without appeal to the presence of a person. Even in the consciousness interpretation, the role of the observer is confined to making the conscious choice of what is to be measured and then unconsciously bringing about what the outcome actually turns out to be. Actuality can only be transformed within the limits of the quantum potentiality already present.

On the neo-Copenhagen view, the experimenter chooses what apparatus to use and so what is to be measured, but then the outcome is decided within that apparatus by macroscopic physical processes. If, on the contrary, it is the new physics of GRW that is at work, it is random process that produces the actual outcome. If Bohmian theory is correct, the role of the observer is simply the classical function of seeing what is already unambiguously the case. In the many-worlds interpretation, it is the observer who is acted upon by physical reality, being cloned to appear in all those parallel universes, within whose vast portfolio all possible outcomes are realized somewhere or other.

No common factor unites these different possible accounts of the role of the observer. At most it would seem appropriate only to speak of 'observer-influenced reality' and to eschew talk of 'observer-created reality'. What was not already in some sense potentially present could never be brought into being.

In connection with this issue, one must also question the assertion, often made in association with claimed parallels to the concept of *maya* in Eastern thought, that the quantum world is a 'dissolving world' of insubstantiality. This is a kind of half-truth. There is the quantum 'veiledness' that we have already discussed, together with the widely acknowledged role that potentiality plays in quantum understanding. Yet there are also persisting aspects of the quantum world that equally need to be taken into account. Physical quantities such as energy and momentum are conserved in quantum theory, much as they are in classical physics. Recall also that one of the initial triumphs of quantum mechanics was to explain the stability of atoms. The quantum exclusion principle undergirds the fixed structure of the periodic table. By no means all the quantum world dissolves into elusiveness.

Quantum hype

It seems appropriate to close this chapter with an intellectual health warning. Quantum theory is certainly strange and surprising, but it is not so odd that according to it 'anything goes'. Of course, no one would actually argue with such crudity, but there is a kind of discourse that can come perilously close to adopting that caricature attitude. One might call it 'quantum hype'. I want to suggest that sobriety is in order when making an appeal to quantum insight.

We have seen that the EPR effect does not offer an explanation of telepathy, for its degree of mutual entanglement is not one that could facilitate the transfer of information. Quantum processes in the brain may possibly have some connection with the existence of the human conscious mind, but random subatomic uncertainty is very different indeed from the exercise of the free will of an agent. Wave/particle duality is a highly surprising and instructive phenomenon, whose seemingly paradoxical character has been resolved for us by the insights of quantum field theory. It does not, however, afford us a licence to indulge in embracing any pair of apparently contradictory notions that take our fancy. Like a powerful drug, quantum theory is wonderful when applied correctly, disastrous when abused and misapplied.

Further reading

Books relating to quantum theory are legion. The following list gives a short personal selection that a reader in search of further insight might find useful to consult.

Books that use more mathematics than this one, while still remaining popular in style:

T. Hey and P. Walters, *The Quantum Universe* (Cambridge University Press, 1987)

J. C. Polkinghorne, *The Quantum World* (Penguin, 1990)

M. Rae, *Quantum Physics: Illusion or Reality?* (Cambridge University Press, 1986)

A book that uses mathematics at a professional level, while being much more concerned with interpretative issues than is usual in textbooks:

C. J. Isham, *Lectures on Quantum Theory: Mathematical and Structural Foundations* (Imperial College Press, 1995)

The classic exposition by one of the founders of the subject:

P. A. M. Dirac, *The Principles of Quantum Mechanics*, 4th edn. (Oxford University Press, 1958)

A philosophically sophisticated discussion of interpretative issues:

B. d'Espagnat, *Reality and the Physicist: Knowledge, Duration and the Quantum World* (Cambridge University Press, 1989)

A more general introduction to issues in the philosophy of science:

W. H. Newton-Smith, *The Rationality of Science* (Routledge and Kegan Paul, 1981)

Newton-Smith, however, neglects the thought of Michael Polanyi, which can be found in:

M. Polanyi, *Personal Knowledge* (Routledge and Kegan Paul, 1958)

Books of special relevance to the Bohmian version of quantum theory:

D. Bohm and B. Hiley, *The Undivided Universe* (Routledge, 1993)
J. T. Cushing, *Quantum Mechanics: Historical Contingency and the Copenhagen Hegemony* (University of Chicago Press, 1994)

Reflective writings by two of the founding figures:

N. Bohr, *Atomic Physics and Human Knowledge* (Wiley, 1958)
W. Heisenberg, *Physics and Philosophy: The Revolution in Modern Science* (Allen & Unwin, 1958)

Biographies of significant quantum physicists:

A. Pais, *Niels Bohr's Times in Physics, Philosophy and Polity* (Oxford University Press, 1991)
H. S. Kragh, *Dirac: A Scientific Biography* (Cambridge University Press, 1990)
A. Pais, *'Subtle is the Lord . . .': The Science and Life of Albert Einstein* (Oxford University Press, 1982)
J. Gleick, *Genius: The Life and Science of Richard Feynman* (Pantheon, 1992)
D. C. Cassidy, *Uncertainty: The Life and Science of Werner Heisenberg* (W. H. Freeman, 1992)
W. Moore, *Schrödinger: Life and Thought* (Cambridge University Press, 1989)

Glossary

Generally speaking, this glossary limits itself to defining terms that recur in the text or that are of particular significance for a basic understanding of quantum theory. Other terms that occur only once or are of less fundamental importance are defined in the text itself, and these can be accessed through the index.

angular momentum: a dynamical quantity that is the measure of rotatory motion

Balmer formula: a simple formula for the frequencies of prominent lines in the hydrogen spectrum

Bell inequalities: conditions that would have to be satisfied in a theory that was strictly local in its character, with no non-local correlations

bosons: particles whose many-particle *wavefunctions* are symmetric

Bohmian theory: a deterministic interpretation of quantum theory proposed by David Bohm

chaos theory: the physics of systems whose extreme sensitivity to details of circumstance makes their future behaviour intrinsically unpredictable

classical physics: deterministic and picturable physical theory of the kind that Isaac Newton discovered

collapse of the wavepacket: the discontinuous change in the *wavefunction* occasioned by an act of measurement

complementarity: the fact, much emphasized by Niels Bohr, that there are distinct and mutually exclusive ways in which a quantum system can be considered

Copenhagen interpretation: a family of interpretations of quantum theory deriving from Niels Bohr and emphasizing indeterminacy and the role of classical measuring apparatus in measurement

decoherence: an environmental effect on quantum systems that is capable of rapidly inducing almost classical behaviour

degrees of freedom: the different independent ways in which a dynamical system can change in the course of its motion

epistemology: philosophical discussion of the significance of what we can know

EPR effect: the counterintuitive consequence that two quantum entities that have interacted with each other retain a power of mutual influence however far apart they may separate from each other

exclusion principle: the condition that no two *fermions* (such as two electrons) can be in the same state

fermions: particles whose many-particle *wavefunctions* are antisymmetric

hidden variables: unobservable quantities that help to fix what actually happens in a deterministic interpretation of quantum theory

interference phenomena: effects arising from the combination of waves, which may result in reinforcement (waves in step) or cancellation (waves out of step)

many-worlds interpretation: an interpretation of quantum theory in which all possible outcomes of measurement are actually realized in different parallel worlds

measurement problem: the contentious issue in the interpretation of quantum theory relating to how one is to understand the obtaining of a definite result on each occasion of measurement

non-commuting: the property that the order of multiplication matters, so that AB is not the same as BA

observables: quantities that can be measured experimentally

ontology: philosophical discussion of the nature of being

Planck's constant: the fundamental new physical constant that sets the scale of quantum theory

positivism: the philosophical position that science is concerned simply with correlating directly observed phenomena

pragmatism: the philosophical position that science is really about the technical capability for getting things done

quantum chaology: the not-fully-understood subject of the quantum mechanics of chaotic systems

quantum field theory: the application of quantum theory to fields such as the electromagnetic field or the field that is associated with electrons

quarks and gluons: current candidates for the basic constituents of nuclear matter

radiation: energy carried by the electromagnetic field

realism: the philosophical position that science is telling us what the physical world is actually like

Schrödinger equation: the fundamental equation of quantum theory that determines how the *wavefunction* varies with time

spin: the intrinsic *angular momentum* possessed by elementary particles

statistical physics: treatment of the bulk behaviour of complex systems on the basis of their most probable states

statistics: the behaviour of systems composed of identical particles

superposition: the fundamental principle of quantum theory that permits the adding together of states that in *classical physics* would be immiscible

uncertainty principle: the fact that in quantum theory *observables* can be grouped in pairs (such as position and momentum, time and energy) such that both members of the pair cannot simultaneously be measured with precise accuracy. The scale of the limit of simultaneous accuracy is set by *Planck's constant*

wavefunction: the most useful mathematical representation of a state in quantum theory. It is a solution of the *Schrödinger equation*

wave/particle duality: the quantum property that entities can sometimes behave in a particlelike way and sometimes in a wavelike way

Mathematical appendix

I set out in concise form, some simple mathematical details that will illuminate, for those who wish to take advantage of them, various points that arise in the mathematically innocent main text. (Items are cross-referenced in that text by their section numbers.) The demands made upon the readers of this appendix vary from the ability to feel at home with algebraic equations to some elementary familiarity with the notation of the calculus.

1. The Balmer formula

It is most helpful to give the formula in the slightly changed form in which it was rewritten by Rydberg. If v_n is the frequency of the nth line in the visible hydrogen spectrum (n taking the integer values, $3, 4, \ldots$), then

$$v_n = cR \left(\frac{1}{2^2} - \frac{1}{n^2} \right), \tag{1.1}$$

where c is the velocity of light and R is a constant called the Rydberg. Expressing the formula in this way, as the difference of two terms, eventually proved to have been an astute move (see section 3 below). Other series of spectral line in which the first term is $1/1^2$, $1/3^2$, etc, were identified later.

2. The photoelectric effect

According to Planck, electromagnetic radiation oscillating v times a second is emitted in quanta of energy hv, where h is Planck's constant and has the tiny value of $6.63.10^{-34}$ joule-seconds. (If one replaces v by the angular frequency $\omega = 2\pi v$, the formula becomes $\hbar\omega$, where $\hbar = h/2\pi$, also often called Planck's constant and pronounced 'aitch bar' or 'aitch slash'.)

Einstein supposed that these quanta had abiding existence. If radiation fell on a metal, one of the electrons in the metal might absorb one quantum, thereby acquiring its energy. If the energy needed for the electron to escape from the metal was W, then that escape would take place if $hv > W$, but it would be impossible if $hv < W$. There was therefore a frequency ($v_0 = W/h$) below which no electrons could be emitted, however intense the beam of incident radiation might be. Above that frequency, some electrons would be emitted, even if the beam were pretty weak.

A pure wave theory of radiation would give an entirely different behaviour, since the energy conveyed to the electrons would then be expected to depend upon the intensity of the beam, but not upon its frequency.

The observed properties of photoelectric emission agree with the predictions of the particle picture and not with the wave picture.

3. The Bohr atom

Bohr supposed that the hydrogen atom consists of an electron of charge $-e$ and mass m moving in a circle around a proton of charge e. The latter's mass is sufficiently large (1,836 times the electron mass) for the effect of its motion to be neglected. If the radius of the circle is r and the electron's velocity is v, then balancing electrostatic attraction against centrifugal acceleration gives

$$\frac{e^2}{r^2} = m\frac{v^2}{r}, \text{ or } e^2 = mv^2r. \tag{3.1}$$

The energy of the electron is the sum of its kinetic energy and electrostatic potential energy, giving

$$E = \frac{1}{2} mv^2 - \frac{e^2}{r},$$ (3.2)

which, using (3.1), can be written as

$$E - \frac{-e^2}{2r}.$$ (3.3)

Bohr then imposed a novel quantum condition, requiring that the angular momentum of the electron must be an integral multiple of Planck's constant $\hbar$,

$$mvr = n\hbar \ (n = 1, 2, \ldots).$$ (3.4)

The corresponding possible energies are then

$$E_n = \frac{-e^4 m}{2\hbar^2} \cdot \frac{1}{n^2}.$$ (3.5)

If the energy released when an electron moves from the state n to the state 2 is emitted as a single photon, the frequency of that photon will be

$$v_n = c \cdot \frac{e^4 m}{4\pi\hbar^3 c} \cdot \left(\frac{1}{2^2} - \frac{1}{n^2} \right).$$ (3.6)

This is just the Balmer formula (1.1). Not only did Bohr explain that formula but he enabled the Rydberg constant R to be calculated in terms of other known physical constants,

$$R = \frac{e^4 m}{4\pi\hbar^3 c},$$ (3.7)

a number that agreed with the experimentally known value. Bohr's discovery represented a remarkable triumph for the new quantum way of thinking

(In the proper quantum mechanical calculation of the hydrogen atom, using the Schrödinger equation (see section 6), the discrete energy levels arise in a somewhat different way, bearing some analogy to the harmonic frequencies of an open string, and the number n is more obliquely related to angular momentum.)

4. Non-commuting operators

The matrices that Heisenberg employed do not in general commute with each other, but eventually it turned out that quantum theory required a further generalization in which non-commuting differential operators were incorporated into the formalism. This is the development that led the physicists eventually to use the mathematics of Hilbert space.

In the general case, quantum mechanical formulae can be obtained from those of classical physics by making the following substitutions for position x and momentum p:

$$x \to x,$$

$$p \to -i\hbar \frac{\partial}{\partial x}. \tag{4.1}$$

Because of the appearance of the differential operator $\partial/\partial x$ in (4.1), the variables x and p do not commute with each other, in contrast to the property of commutation that trivially applies to the numbers that classical physics assigns to positions and momenta. When $\partial/\partial x$ is on the left it differentiates the x on its right, as well as any other entity to the right, so that we may write

$$\frac{\partial}{\partial x} \cdot x - x \cdot \frac{\partial}{\partial x} = 1. \tag{4.2}$$

Defining the commutator bracket $[p,x] = p.x - x.p$, we may rewrite this as

$$[p,x] = -i\hbar. \tag{4.3}$$

This relation is known as the *quantization condition*. An alert reader will note that another solution of (4.3) would be given by

$$x \to i\hbar \frac{\partial}{\partial p},$$

$$p \to p. \tag{4.4}$$

Dirac particularly emphasized the way in which there are many equivalent ways of formulating quantum mechanics.

5. de Broglie waves

The Planck formula

$$E = h\nu \tag{5.1}$$

makes energy proportional to the number of vibrations per unit interval of *time*. Relativity theory brackets together space-and-time, momentum-and-energy, as natural fourfold combinations. The young de Broglie therefore proposed that in quantum theory momentum should be proportional to the number of vibrations per unit interval of *space*. This leads to the formula

$$p = \frac{h}{\lambda}, \tag{5.2}$$

where λ is the wavelength. Equations (5.1) and (5.2) together give a way of relating particlelike properties (E and p) to wavelike properties (ν and λ). The spatial dependence of a waveform of wavelength λ is given by

$$e^{i2\pi x/\lambda} \tag{5.3}$$

Putting (4.1) and (5.3) together recovers (5.2).

6. The Schrödinger equation

The energy of a particle is the sum of its kinetic energy ($\frac{1}{2}mv^2 = \frac{1}{2}p^2/m$, where p is mv) and its potential energy (which, in general, we can write as a function of x, $V(x)$). The quantum mechanical relation between energy and time that is the analogue of (4.1) is

$$E \rightarrow i\hbar \frac{\partial}{\partial t}. \tag{6.1}$$

The difference in sign between (6.1) and (4.1) is due to the fact that the time dependence of a waveform moving to the right and corresponding to the spatial dependence (5.3), is

$$e^{-i2\pi vt}, \tag{6.2}$$

so that the plus sign in (6.1) is needed to give $E = hv$.

Using (4.1) and (6.1) to turn $E = \frac{1}{2}mv^2 + V$ into a differential equation for the quantum mechanical wavefunction ψ, yields

$$i\hbar \frac{\partial \psi}{\partial t} = \left[-\frac{\hbar^2}{2m} \frac{\partial^2}{\partial x^2} + V(x) \right] \psi, \tag{6.3a}$$

in one space dimension, and

$$i\hbar \frac{\partial \psi}{\partial t} = \left[-\frac{\hbar^2}{2m} \nabla^2 + V(\boldsymbol{x}) \right] \psi, \tag{6.3b}$$

in the three-dimensional space of the vector $\boldsymbol{x} = (x, y, z)$, where

$$\nabla^2 = \frac{\partial^2}{\partial x^2} + \frac{\partial^2}{\partial y^2} + \frac{\partial^2}{\partial z^2}. \tag{6.4}$$

These expression are the Schrödinger equation, first written down by him on the basis of a rather different line of argument. The operator in square brackets in equations (6.3) is called the *Hamiltonian*.

Notice that equations (6.3) are *linear equations* in ψ, that is to say if ψ_1 and ψ_2 are two solutions, so is

$$\lambda_1\psi_1 + \lambda_2\psi_2, \tag{6.5}$$

for any pair of numbers λ_1 and λ_2.

Max Born emphasized that the wavefunction has the interpretation of representing a probability wave. The probability of finding a particle at the point x is proportional to the square of the modulus of the corresponding (complex) wavefunction.

7. Linear spaces

The linearity property noted at the end of section 6 is a fundamental characteristic of quantum theory and the basis for the superposition principle. Dirac generalized the ideas based on wavefunctions, formulating the theory in terms of abstract vector spaces.

A set of vectors $|a_i\rangle$ form a *vector space* if any combination of them

$$\lambda_1|a_1\rangle + \lambda_2|a_2\rangle + \ldots, \tag{7.1}$$

also belongs to the space, where the λ_i are arbitrary (complex) numbers. Dirac called these vectors 'kets'. They are the generalizations of the Schrödinger wavefunctions ψ. There is also a dual space of 'bras', antilinearly related to the kets

$$\sum_i \lambda_i|a_i\rangle \rightarrow \sum_i \langle a_i|\lambda_i^* \tag{7.2}$$

where the λ_i^* are the complex conjugates of the λ_i. (The bras $\langle a|$

obviously correspond to the complex conjugate wavefunctions, ψ^*.) A scalar product can be formed between a bra and a ket (giving a 'bra(c)ket' – Dirac was rather fond of this little joke). This corresponds in wavefunction terms to the integral $\int \psi_1^* \psi_2 \mathrm{d}x$. It is denoted by $\langle a_1 | a_2 \rangle$ and it has the property that

$$\langle a_1 | a_2 \rangle = \langle a_2 | a_1 \rangle^*. \tag{7.3}$$

It follows from (7.3) that $\langle a | a \rangle$ is a real number and, in fact, in quantum theory the condition is imposed that it is positive (it must correspond to $| \psi |^2$).

The relationship between a physical state and a ket is what is called a *ray representation*, meaning that $| a \rangle$ and $\lambda | a \rangle$ represent the same physical state for any non-zero complex number λ.

8. Eigenvectors and eigenvalues

Operators on vector spaces are defined by their effect of turning kets into other kets:

$$O | a \rangle = | a' \rangle. \tag{8.1}$$

In quantum theory, operators are the way in which observable quantities are represented in the formalism (compare with the operators (4.1) acting on a wavefunction). Significant expressions are the numbers that arise as bra-operator-ket 'sandwiches' (called 'matrix elements'; they are related to probability amplitudes):

$$\langle \beta | O | a \rangle. \tag{8.2}$$

The *hermitean conjugate* of an operator, $O^\dagger$, is defined by the relation between matrix elements:

$$\langle \beta | O | a \rangle = \langle a | O^\dagger | \beta \rangle^*. \tag{8.3}$$

Special significance attaches to operators that are their own hermitean conjugate:

$$O^\dagger = O. \hspace{4cm} (8.4)$$

They are called *hermitean*, and only such operators represent physically observable quantities.

Since the results of actual observations are always real numbers, to make physical sense of this scheme there has to be a way of associating numbers with operators. This is established using the ideas of *eigenvectors* and *eigenvalues*. If an operator O turns a ket $|a\rangle$ into a numerical multiple of itself,

$$O|a\rangle = \lambda|a\rangle, \hspace{4cm} (8.5)$$

then $|a\rangle$ is said to be an eigenvector of O with eigenvalue λ. It can be shown that the eigenvalues of hermitean operators are always real numbers.

The physical interpretation corresponding to these mathematical facts is that the real eigenvalues of an observable are the possible results that can be obtained by measuring that observable, and the associated eigenvectors correspond to the physical states in which that particular result will be obtained with certainty (probability one). Only two observables whose corresponding operators commute will be simultaneously measurable.

9. The uncertainty relations

The discussion of the gamma-ray microscope has shown that quantum measurement forces on the observer some trade-off between good spatial resolution (short wavelength) and small disturbance (low frequency). Putting this balance into quantitative terms leads to the Heisenberg uncertainty relations, where it is found that the uncertainty in position, Δx, and the uncertainty in momentum, Δp, cannot have a product $\Delta x \cdot \Delta p$ whose magnitude is less than the order of Planck's constant $\hbar$.

10. Schrödinger and Heisenberg

If H is the Hamiltonian (energy operator), Schrödinger's equation reads,

$$i\hbar \frac{\partial |a, t\rangle}{\partial t} = H |a, t\rangle. \tag{10.1}$$

If H does not depend explicitly on time, as is usually the case, (10.1) can be solved formally by writing

$$|a, t\rangle = e^{-iHt/\hbar} |a, 0\rangle. \tag{10.2}$$

The physical consequences of the theory all derive from the properties of matrix elements of the form $\langle a | O | \beta \rangle$. Writing out explicitly the time-dependence (10.2) gives

$$\langle a, 0 | e^{iHt/\hbar} . O . e^{-iHt/\hbar} | \beta, 0 \rangle. \tag{10.3}$$

Associating the terms together in a different way gives

$$\langle a, 0 | . e^{iHt/\hbar} O e^{-iHt/\hbar} . | \beta, 0 \rangle \tag{10.4}$$

where now the time-dependence has been thrust, so to speak, onto a time-dependent operator

$$O(t) = e^{iHt/\hbar} O e^{-iHt/\hbar} \tag{10.5}$$

(10.5) can then be treated as the solution of the differential equation

$$i\hbar \frac{\partial O(t)}{\partial t} = OH - HO = [O, H]. \tag{10.6}$$

This way of thinking about quantum theory, in which time-dependence is associated with the operator observables rather than with the states, is just the way in which Heisenberg originally approached the matter. Thus the discussion of the section has demonstrated the equivalence of the approaches of the two great founding figures of quantum theory, despite the appearance initially of their having treated the question so very differently.

11. Statistics

If 1 and 2 are identical and indistinguishable particles, then $|1, 2\rangle$ and $|2, 1\rangle$ must correspond to the same physical state. Because of the ray representation character of the formalism (see section 7), this implies that

$$|2, 1\rangle = \lambda |1, 2\rangle \tag{11.1}$$

where λ is the number. However, interchanging 1 and 2 twice is no change at all and so it must restore exactly the original situation. Therefore it must be the case that

$$\lambda^2 = 1, \tag{11.2}$$

giving the two possibilities, $\lambda = +1$ (bose statistics), or $\lambda = -1$ (fermi statistics).

12. The Dirac equation

On the memorial tablet to Paul Dirac in Westminster Abbey, there is engraved the equation:

$$i\gamma\partial\psi = m\psi. \tag{12.1}$$

It is his celebrated relativistic wave-equation for the electron, written in a four-dimensional spacetime notation and (using the physical units natural to quantum theory that set $\hbar = 1$). The γs are 4 by 4 matrices and ψ is what is called a four-component spinor (2 (spin) times 2 (electron/positron) states). That is as far as we can take the matter in an introductory book of this kind, but, whether on paper, on the page, or on stone in the Abbey, the onlooker should have the opportunity to pay respect to what is one of the most beautiful and profound equations in physics.

Index

Visit the
VERY SHORT INTRODUCTIONS
Web site

www.oup.co.uk/vsi

➤ **Information** about all published titles

➤ News of **forthcoming books**

➤ **Extracts** from the books, including titles not yet published

➤ **Reviews** and views

➤ **Links** to other **web sites** and main OUP web page

➤ Information about **VSIs in translation**

➤ **Contact** the editors

➤ **Order** other **VSIs** on-line